# JOLPIC MATH!

## Easy to learn

# ALGEBRA

## for Kids and Beginners

# JOLPIC KIDZ

# JOLPIC MATH!
# EASY TO LEARN ALGEBRA
# FOR KIDS AND BEGINNERS

**First Published: January 2024**

Jolpic Kidz is a publishing company of educational books written for kids.

For more info, mail us at jolpic@gmail.com

# CONTENTS

**CHAPTER 1**
Know about numbers **4**

**CHAPTER 2**
Four Simple Operations: Addition, **52**
Subtraction, Multiplication, and
Division

**CHAPTER 3**
Representation of Numbers by letters **90**

**CHAPTER 4**
Some Useful Relations and Formulas **104**

**CHAPTER 5**
Indices or Powers **128**

**CHAPTER 6**
Equations and Their Solutions **142**

**CHAPTER 7**
Factorization **154**

# CHAPTER 1

# KNOW ABOUT NUMBERS

In our everyday life, we use numbers to count things or measure quantities. These numbers are represented by some symbols, 0, 1, 2, 3, 4, 5, 6, 7, 8, and 9. Using these symbols, we are able to write from any small number to any large number.

0 1 2 3
4 5 6
7 8 9

**But if we would not use these numbers, could we count things? Let's try to find out.**

Suppose you own some books, but you do not know how many books you have. You need to count them to determine the exact number of your books. But the condition is that you should not use any number in order to count them. Then how do you count?

**Take a pen and a piece of paper to try it yourself.**

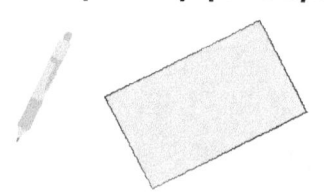

You visualize the first book, and draw a line on the paper.

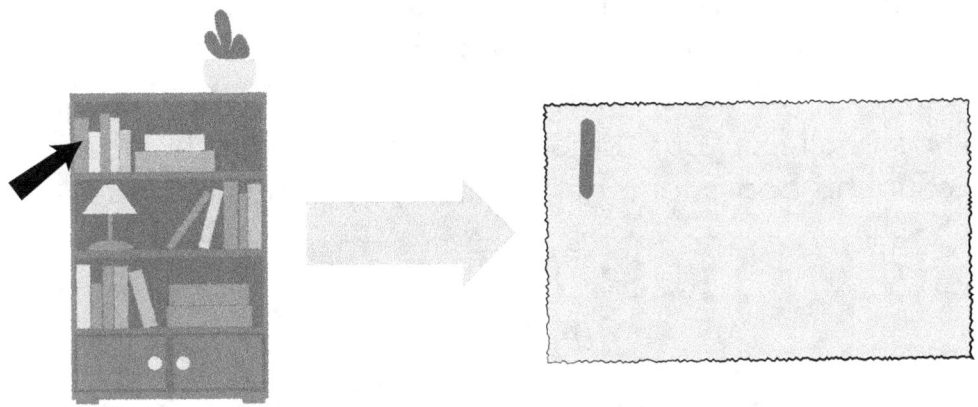

Then you visualize the next book, and you draw another line beside the previous line.

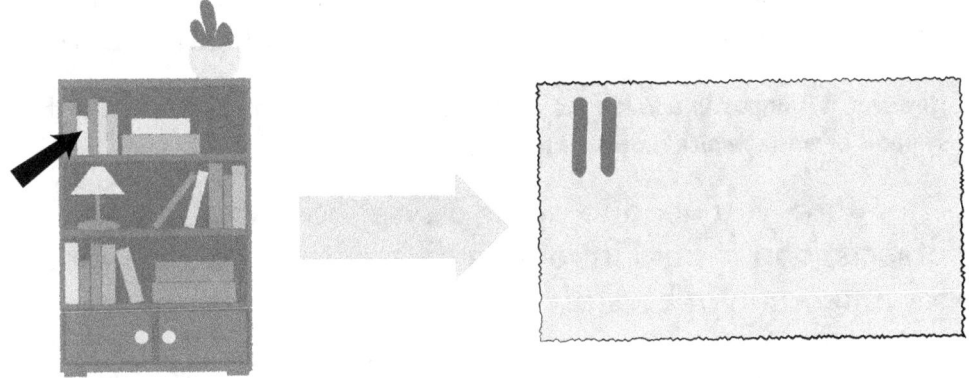

That way you get several lines on the paper when you finish your counting.

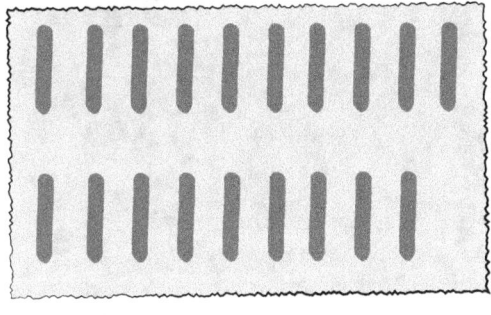

However you will be not able to tell how many books you have until you count the total number of lines on the paper by the conventional method.

You merely get an idea whether you own a satisfactory number of books or not. Also you can compare the number of your books with your friend.

Suppose you have a friend named Bob, who has a similar bookshelf like you. You employed the same technique to count his books, and get some lines on another paper.

### Your books        Bob's books

Now you can visually compare these lines on both papers, and tell you have a greater number of books than Bob.

**However, it is impossible to tell the exact amount of books until you introduce the symbols or words, which represent numbers.**

Therefore, you need to apply a systematic way to count objects, and should introduce different symbols or words for different numbers of lines.

| Lines on the paper | Symbols | Words |
|---|---|---|
| I | 1 | One |
| II | 2 | Two |
| III | 3 | Three |

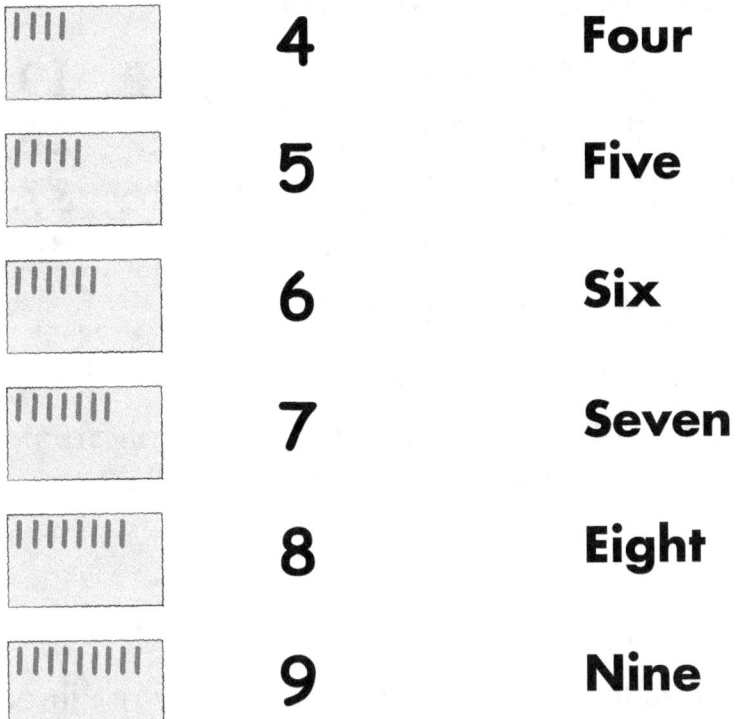

| | 4 | Four |
| | 5 | Five |
| | 6 | Six |
| | 7 | Seven |
| | 8 | Eight |
| | 9 | Nine |

If we continued to introduce a new symbol in each increment of the number of lines, we would find a great difficulty. We needed a unique symbol for each number. But you know that a number can be very large. Therefore, the use of a unique symbol for every possible number is not a good idea, because one would have to remember an infinite number of symbols to represent any arbitrary number.

However, we have a solution to that problem. Instead of using a unique symbol for each number, we should think of a base. In the traditional numbering system, we use base 10.

When you find the number of lines on the paper is one line greater than 9, you should write 1 and a 0 (zero).  **10 (Ten)**

**You already know how to express other two digits numbers.**

11, 12, 13, 14, 15, 16, 17, 18, 19,
20, 21, 22, 23, 24, 25, 26, 27, 28,
29, 30, 31, 32, 33, 34, 35, 36, 37,
38, 39, 40, 41, 42, 43, 44, 45, 46,
47, 48, 49, 50, 51, 52, ..............

Following the way, you can write numbers up to 99.

If you want to write a number which is one unit greater than 99, you have to write the digit 1 followed by two 0.

# 100

This number is called one hundred.

If you continue that way, you will be able to write three digits numbers from 100 to 999. After 999, you have to introduce a four digits number 1000, which is called one thousand. And the process will go on.

This method of numbering allows you to demonstrate any number with the help of these ten digits, 0, 1, 2, 3, 4, 5, 6, 7, 8, 9.

This number system is known as the decimal number system, because here we use base 10. Apart from this numbering system, other systems of numbering, such as octal, binary, hexadecimal, are also used in some specific fields of mathematics.

The decimal numbering system is very easy to handle. Applying this numbering system, we can write any range of numbers just by increasing the number of digits. But first of all, we need to understand the meaning of place value.

| | |
|---|---|
| 1 | One |
| 10 | Ten |
| 100 | Hundred |
| 1000 | Thousand |
| 10000 | Ten thousand |
| 100000 | Hundred thousand |
| 1000000 | Million |
| 10000000 | Ten million |
| 100000000 | Hundred million |
| 1000000000 | Billion |

Now consider an arbitrary large number, and check how big it is.

# 9743

The place values of its digits are,

| Thousands | Hundreds | Tens | Ones |
|---|---|---|---|
| 9 | 7 | 4 | 3 |

Therefore, the number can be written in words also,

**Nine thousand seven hundred forty three**

You can also split the number, 9743, in the following way.

The meaning of the above is, 9743 is the sum of,

9 times thousand
7 times hundred
4 times ten
3 times one

$$9743 = 9 \times 1000 + 7 \times 100 + 4 \times 10 + 3$$

**You can treat any large number in this way.**

## Natural Numbers

Any whole number excluding 0 (zero) are known as natural numbers. Using natural numbers, you can count any number of physical things.

The natural numbers are,

**1, 2, 3, 4, 5, 6, 7, 8, 9, 10, 11, 12, 13, 14, 15....**

## Significance of Zero

We often use the number 0 (zero) to signify nothing. For example, someone asked you, "How many books are there on the table?"

As you can see there is no book on the table.
So your answer should be, "There is no book on the table."

You can alternatively say,
"There is 0 (zero) book on the table."

## Even Numbers and Odd Numbers

If a number is exactly divisible by 2, the number is called an even number.

For example, 2, 4, 6, 8, 10, 12, etc. are even numbers because these are exactly divisible by 2.

On the other hand, if a number is not divisible by 2, the number is called an odd number.

1, 3, 5, 7, 9, 11, 13, etc. are odd numbers because these are not divisible by 2.

## Prime numbers

Most of the natural numbers are multiple of smaller whole numbers.

For example, 6 is the multiple of 2 and 3.

$$6 = 2 \times 3$$

8 is the multiple of 2 and 4.

$$8 = 2 \times 4$$

24 is the multiple of 2, 3, 4, 6, 8, and 12.

| | |
|---|---|
| 24 = 2 × 12 | 24 = 6 × 4 |
| 24 = 3 × 8 | 24 = 8 × 3 |
| 24 = 4 × 6 | 24 = 12 × 2 |

Also, all natural numbers are multiple of 1 and the number itself.

For example, 6 is also divisible by 1 and 6 along with 2 and 3.

$$6 = 1 \times 6$$
$$6 = 6 \times 1$$

8 is divisible by 1 and 8 also along with 2 and 4.

$$8 = 1 \times 8$$
$$8 = 8 \times 1$$

24 is also divisible by 1 and 24 along with 2, 3, 4, 6, 8, and 12.

$$24 = 1 \times 24$$
$$24 = 24 \times 1$$

But there are some numbers that are only divisible by 1 and that number. They are not divisible by any other smaller number. Those numbers are termed as prime numbers.

For example, 7 is the multiple of 1 and 7, thus 7 is a prime number.

$$7 = 1 \times 7$$
$$7 = 7 \times 1$$

Similarly, 19 is divisible by 1 and 19 only, thus 19 is also a prime number.

$$19 = 1 \times 19$$
$$19 = 19 \times 1$$

- 2 is considered as a prime number because it is the multiple of 1 and the number itself.

- 2 is the smallest prime number, and it is the only possible even prime number.

- All other even numbers are always non-prime because they are divisible by 2.

- List of prime numbers from 1 to 100
  2, 3, 5, 7, 11, 13, 17, 19, 23, 29, 31, 37, 41, 43, 47, 53, 59, 61, 67, 71, 73, 79, 83, 89, 97.

- 1 is not considered as a prime number.

- Prime numbers become less frequent as numbers get bigger. For example, there are 25 prime numbers 1 to 100. But you will get only 14 prime numbers between the range 901 - 1000.

- There exist endless numbers of prime numbers, and mathematicians are still trying to find new prime numbers, which may be very large numbers.

- It is quite difficult to predict a prime number because there is not any known formula, which could tell the series of all prime numbers one by one.

## Fractions

So far our discussions were limited to whole numbers. We are able to count any number of objects with the help of those numbers.

**Examples:**

How many fingers do you have in your both hands?

Your answer should be 10, which is a whole number.

How many planets are there revolving around the sun?

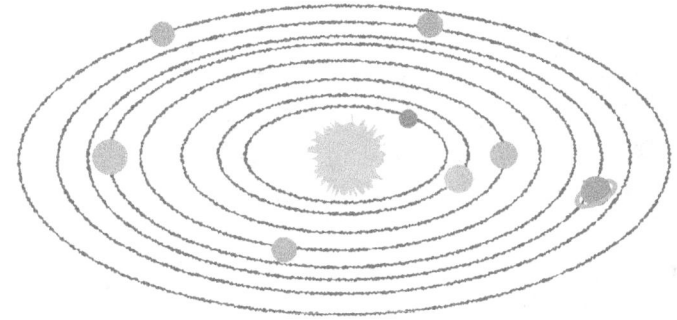

Your answer will be 8, which is also a whole number.

That is the way you can tell the number of any collection of objects using whole numbers. You can count the number of pages of any book, number of fishes in an aquarium, number of teeth inside your mouth, and so on.

Besides whole numbers, fractions are also used in mathematics. We often use fractions when a whole number fails to describe the exact amount of something. We should understand the concept of fractional numbers with the help of a few examples.

13

Suppose your father gifted you 3 chocolate bars on your birthday.

You could have eaten all of them yourself.

But being a wise person, you decided to share your happiness with your friend, Bob. So you ate 2 chocolates whole, and offered some portions of the third one to Bob as shown in the figure.

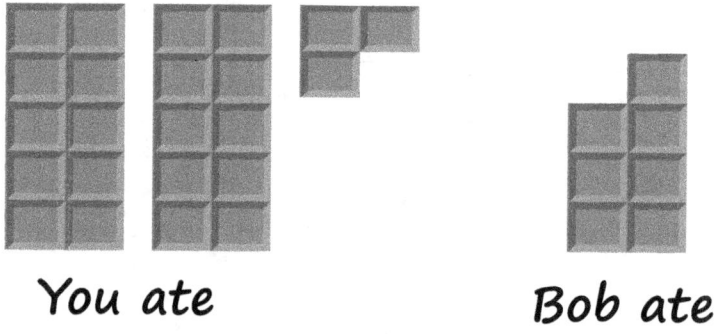

## You ate          Bob ate

As you notice, the chocolate bar is divided into 10 equal parts, and you gave 7 of them to Bob, and then ate the remaining 3 parts yourself.

**Now answer the following questions:**

**How many chocolates did you eat?**
**How many chocolates did Bob eat?**

It is an obvious fact that you did not eat 3 whole chocolates. So your answer will not be 3. Also, you eat 2 whole chocolates, and some portions of the third one. So, your answer will not be 2 either.

Since the amount of chocolate is more than 2 but less than 3, the answer should be a number between 2 and 3.

In order to express a number, whose value lies between two whole numbers, you need to understand the concept of fractions.

As you see in the figure, 10 parts of a chocolate represent one whole chocolate. Therefore, each part of the chocolate can be represented by $\frac{1}{10}$.

**1 chocolate**

**$\frac{1}{10}$ chocolate**

Since you ate 3 pieces among the ten for the third chocolate, you may say that you ate $\frac{3}{10}$ chocolate in case of the third chocolate bar.

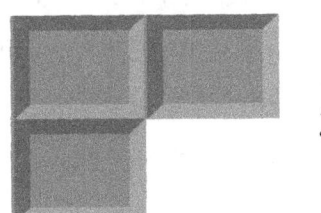

**$\frac{3}{10}$**

This $\frac{3}{10}$ is a fractional number. You can also write this fraction as,

**0.3**

Therefore, total number of chocolates that you ate is,

$$1 + 1 + \frac{3}{10}$$

$$= 2\frac{3}{10} \text{ chocolates}$$

You can alternatively write,

$$1 + 1 + 0.3 = 2.3 \text{ chocolates}$$

Likewise, you can calculate the amount of chocolate that was eaten by Bob,

**$\frac{7}{10}$ or 0.7 chocolate.**

Numbers are not only used for counting physical objects, but also for measuring several quantities. While we measure some quantities, such as length, weight, time, etc., we encounter fractions in most of the cases.

For example, if you are measuring the height and the width of a credit card using a simple ruler, you will find the height and the width of the card are 5.398 cm and 8.56 cm respectively. These two numbers are fractions. Similarly, when you measure the weights of several objects using a digital balance, you will see the reading on the screen is in fractional numbers most of the time.

## How to represent fractions

We just discussed how we can mathematically represent a fraction. A fraction can be either represented by the ratio of two whole numbers (such as ½), or in decimal form (such as 0.5).

½ and 0.5 have the equal values. It means they are actually the same number represented in different forms. However, it is not very hard to express a given fraction in either of these two forms. We should learn the method with the given example.

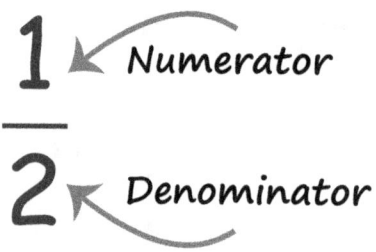

This fraction is a ratio of two whole numbers, 1 and 2.

The number written above the line is called numerator, and the number written below the line is called denominator.

**The meaning of the above fraction is division of the numerator by its denominator.**

We can do the calculation in the following way,

$$\frac{Numerator}{Denominator}$$

The above expression is equivalent to

**Numerator ÷ Denominator**

Therefore, ½ means

$$1 \div 2$$

Here 1 is less than 2, so you can not divide 1 by 2 applying the simple method of division. You need to apply a special technique here.

$$2 \overline{)\ 1}$$

You will be able to add a 0 at the right hand side of the dividend (here the dividend is 1) after writing 0. in the quotient box.

$$2 \overline{)\ 10} \quad 0.$$

Now the dividend becomes 10, which is larger than the divisor 2. Thus it is possible to divide 10 by 2, and the final answer becomes 0.5.

$$\begin{array}{r} 0.5 \\ 2 \overline{)\ 10} \\ 10 \\ \hline \times \end{array}$$

$$\frac{1}{2} \quad \Longrightarrow \quad 0.5$$

*Decimal point*

It is also possible to show a decimal fraction expressed into the ratio of two whole numbers.
We start with the same decimal fraction that we obtained from the previous example.

$$0.5$$

Our primary aim is to determine the numerator and the denominator of the fraction of its ratio form.

The numerator of the fraction is the number written at the right hand side of the decimal point. In this case, the number written at the right hand side of the decimal is 5, thus the numerator will be 5.

## Numerator = 5

The denominator of the fraction will be one of the following numbers.

## 10, 100, 1000, 10000, 100000...

In the present case, we should choose 10 as the denominator, because there is one digit after the decimal point. The number of digits must be equal to the number of 0 written after 1.

## Denominator = 10

Therefore the expression of the fraction should be,

$$\frac{5}{10}$$

Since the expression is a ratio of two numbers, we can cancel out the common factors of them.

$$\frac{5}{10} = \frac{1 \times 5}{2 \times 5} = \frac{1}{2} \qquad 0.5 \implies \frac{1}{2}$$

## Another Example

### Fraction to decimal

Consider the fraction, $\dfrac{3}{8}$

We are going to transform the fraction into its decimal form.

$$
\begin{array}{r}
0.375 \\
8\overline{)\,3.000} \\
24\phantom{00} \\
\hline
60\phantom{0} \\
56\phantom{0} \\
\hline
40 \\
40 \\
\hline
0
\end{array}
$$

Therefore the decimal form of the fraction is 0.375.

### Decimal to fraction

Consider the decimal, **0.375**

Here the numerator should be 375 because the number written after decimal point is 375.

And the denominator should be 1000 because there are 3 digits after the decimal point.

Therefore the corresponding fractional form should be,

$$\frac{375}{1000} = \frac{3 \times 125}{8 \times 125} = \frac{3}{8}$$

## Repeating Decimals

Some fractions are quite interesting. When any of such fractions is represented in its decimal form, a never-ending repetition of the digits on the right hand side of the decimal point is observed.

To demonstrate, we consider the fraction, $\dfrac{1}{3}$

Let's see what happens when we try to transform it into the decimal form.

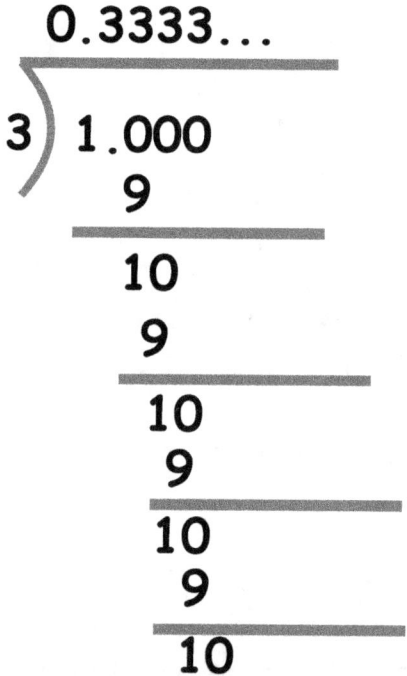

The division will continue that way, and there will always be a non zero remainder in every step.

The decimal form of this fraction is,

**0.33333333333333...**

Therefore, the decimal form of the fraction, $^1/_3$ is a repeating decimal.

Repeating decimals are often represented with a bar (-) sign.

For example, the decimal form of $^1/_3$ is 0.3333333....,
which can be alternatively represented as,

$$0.\overline{3}$$

The bar sign written above the digit implies that the digit (or digits) infinitely repeats itself.

$$\frac{1}{9} = 0.1111111111... = 0.\overline{1}$$

$$\frac{5}{33} = 0.151515151515... = 0.\overline{15}$$

$$\frac{3}{11} = 0.27272727272727... = 0.\overline{27}$$

$$\frac{2}{3} = 0.666666666... = 0.\overline{6}$$

If a repeating decimal is given, it can be represented as the ratio of two whole numbers too. But this time, the procedure is a bit different from the previous one.

Consider the following repeating decimal is given to you.

$$0.18181818181818...$$

Here you see that at the right hand side of the decimal point, the two digit number, 18 is repeating itself forever. Therefore, you can also write the decimal fraction as,

$$0.\overline{18}$$

Now we have to determine the numerator and the denominator of that fraction.

In such cases, the numerator will be the repeating number at the right hand side of the decimal point. Here the number 18 is repeating, thus the numerator will be 18.

## Numerator = 18

The possible denominators of repeating decimal fractions are,

## 9, 99, 999, 9999, 99999, 999999...

Note here that the possible denominators should not be 10, 100, 1000, 10000, etc., because a repeating decimal is concerned.

For repeating decimal, the number of digits of the numerator and the denominator will be equal. Since the numerator is 18, and it is a two digits number, we should take 99 as the denominator, which is also a two digits number.

## Denominator = 99

Therefore the fraction becomes,

$$\frac{18}{99}$$

Now we should divide the numerator and the denominator by the highest common factor of them to make the fraction simple.

$$\frac{18}{99} = \frac{2 \times 9}{11 \times 9} = \frac{2}{11}$$

We may show more examples of repeating decimal fractions, and express them as the ratio of two whole numbers.

**Example:**

$$0.777777777...$$

The alternative expression of the above repeating decimal is,

$$0.\overline{7}$$

Here the digit 7 is repeating, so the numerator of the fraction will be 7.

## Numerator = 7

Since 7 is a one digit number, the appropriate denominator will be 9.

## Denominator = 9

Therefore the final expression of the fraction is,

$$\frac{7}{9}$$

**Example:**

$$0.07070707070707...$$

The given repeating decimal can be expressed as,

$$0.\overline{07}$$

Here the two digits, 0 and 7 are repeating after the decimal point. In order to determine the numerator, we should neglect 0, as it is written on the left hand side of the number 7. Therefore, the numerator will be 7.

## Numerator = 7

However, we should consider 07 as a two digits number while we determine the denominator. Thus the denominator will be 99.

## Denominator = 99

Therefore, the final expression of the fraction is,

$$\frac{7}{99}$$

**Example:**

$$0.142857142857142857142857...$$

The above repeating decimal looks complicated, but at the end, the final expression is going to be very simple. Let's try to express it as the ratio of two whole numbers.

Here we see the six digit number, 142857 is repeating on the right hand side of the decimal point. Thus the alternative expression is,

$$0.\overline{142857}$$

Thus the numerator of the corresponding fraction should be,

## Numerator = 142857

Since the repeating number is a six digit number, the denominator should be,

## Denominator = 999999

Therefore, the corresponding fraction is,

$$\frac{142857}{999999}$$

The highest common factor of the numerator and the denominator is 142857. Therefore, the final expression of the fraction we get,

$$\frac{142857}{999999} = \frac{1 \times 142857}{7 \times 142857} = \frac{1}{7}$$

Following this method, it is possible to express most of the repeating decimals into their ratio form. However for some cases, the method does not work.

For example, you cannot apply this method to the following repeating decimal

### 0.83333333333...

Notice here the number 3 is repeating infinitely, but there is an extra digit 8, which is not repeating.

The corresponding ratio form of the decimal fraction is,

$$\frac{5}{6}$$

But, the previous technique is not sufficient to get the above expression. Later we will see how algebra helps us to determine the fraction forms of such repeating decimals.

There are more examples of such decimals.

$$0.916666666... = \frac{11}{12}$$

$$0.58333333... = \frac{7}{12}$$

$$0.16666666... = \frac{1}{6}$$

## Rational Numbers and Irrational Numbers

Every fraction (represented as a ratio of two whole numbers) yields either a terminating decimal or a repeating decimal.

Examples:

$$\frac{1}{10} = 0.1 \longrightarrow \textit{Terminating decimal}$$

$$\frac{1}{9} = 0.1111111\ldots \longrightarrow \textit{Repeating decimal}$$

Therefore whenever you see a terminating decimal, or a repeating decimal, it is confirmed that the decimal can also be represented as the ratio of two whole numbers.

This fact is also true for any whole number. You are able to represent a whole number as the ratio of two whole numbers.

For example, 2 is a whole number, and you can represent it as,

$$2 = \frac{2}{1} \qquad 2 = \frac{6}{3} \qquad 2 = \frac{8}{4}$$

and several other ratios of whole numbers.

Similarly, 5 is a whole number, and it can be represented as,

$$5 = \frac{10}{2} \qquad 5 = \frac{15}{3} \qquad 5 = \frac{20}{4}$$

Therefore, all whole numbers and the mentioned decimal fractions have the property that each of them can be represented as the ratio of a pair of whole numbers.

When a number, maybe a whole number or a fraction, can be represented as the ratio of two whole numbers, the number is known as a rational number.

Apart from these numbers, there are several decimal fractions that cannot be represented as the ratio of two whole numbers. These numbers are termed as irrational numbers.

## Examples of Irrational Numbers

### Pi ($\pi$)

The ratio of circumference to diameter of a circle yields an unique number called pi ($\pi$). A circle may be large or small, or may have any size, but the ratio of circumference to diameter for any circle always gives a constant value.

The value of pi is,

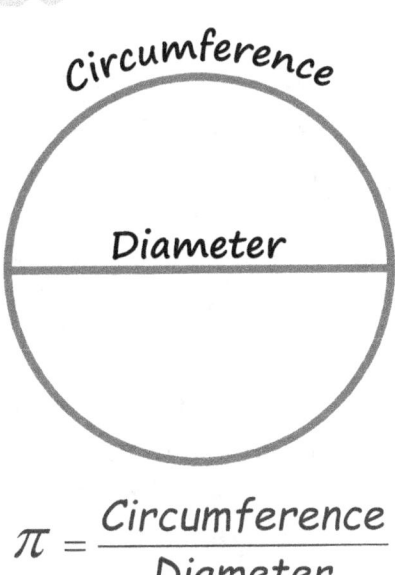

$$\pi = \frac{Circumference}{Diameter}$$

3.141592653589793238...

This value is denoted by the symbol, $\pi$.

$$\pi$$

This number is a nonterminating decimal because there are an infinite number of digits on the right hand side of the decimal point, but they are not repeating. Due to the reason, the number, π (pi) can not be represented as the ratio of two whole numbers. Therefore, π is an irrational number.

# Square roots of imperfect squares

**Square roots of imperfect squares are also irrational numbers.**

If a number is multiplied by itself, the product is called the square of that number.

For example,

$$2 \times 2 = 4 \quad \text{or} \quad 2^2 = 4$$

Square of 2 is 4

Similarly,
$$3^2 = 3 \times 3 = 9$$
$$5^2 = 5 \times 5 = 25$$
$$6^2 = 6 \times 6 = 36$$

Square root is just the reverse operation of square, and generally it is expressed by the symbol,

$$\sqrt{\phantom{x}}$$

If you consider the above examples, you may say,

2 is the square root of 4
3 is the square root of 9
4 is the square root of 16
5 is the square root of 25
6 is the square root of 36

Alternatively, you can express these statements mathematically as,

$$\sqrt{4} = 2 \qquad \sqrt{25} = 5$$

$$\sqrt{9} = 3 \qquad \sqrt{36} = 6$$

$$\sqrt{16} = 4 \qquad \sqrt{49} = 7$$

Here 4, 9, 16, 25, 36, 49, etc., are called perfect squares because the square roots of these numbers are whole numbers.

But most of the whole numbers are imperfect squares, because the square roots of those numbers are not whole numbers, but decimals.

For example, 2, 3, 5, 7, etc., are imperfect squares as their square roots are not whole numbers.

$$\sqrt{2} = 1.41421356237...$$

$$\sqrt{3} = 1.7320508075...$$

$$\sqrt{5} = 2.236067977...$$

$$\sqrt{7} = 2.6457513110...$$

These square roots are non terminating decimals, but their digits are not repeating. Therefore, these are also irrational numbers, and they cannot be expressed as the ratio of two whole numbers.

**Therefore, as per the definition of irrational numbers, those decimals are also irrational numbers.**

## Proper, Improper, and Mixed Fractions

**Fractions are divided into three categories, proper, improper, and mixed fractions. It is time to talk about them.**

## Proper Fractions

When the numerator of a fraction is less than its denominator, the fraction is called a proper fractions.

**Examples:**

$$\frac{1}{2}, \frac{2}{3}, \frac{3}{11}, \frac{4}{19}, \frac{4}{7}$$

*These fractions are proper fractions.*

**A proper fraction is always less than 1.**

To understand the statement, we need to recall the example of chocolate pieces.

A whole chocolate bar can be divided into some equal parts. In the present case, the chocolate bar has been divided into 10 equal parts.

Each part of the chocolate is expressed by the fraction, $^1/_{10}$.

$\dfrac{1}{10}$

In a similar way, you may represent 3 parts of the chocolate as $^3/_{10}$, or 7 parts of the chocolate as $^7/_{10}$, and so on.

$$\frac{3}{10}\qquad\qquad \frac{7}{10}$$

When you consider all the 10 parts, they represent the whole number, 1 ($^{10}/_{10} = 1$), which is the whole chocolate bar.

$$\frac{10}{10} = 1$$

Here you can see when the numerator is less than the denominator, the corresponding fraction represents some portions of the chocolate bar, instead of the whole chocolate bar. But when the numerator becomes equal to the denominator, the value becomes 1, which is a whole number, and it represents the entire chocolate bar.

**Therefore, a proper fraction, which has a larger denominator and smaller numerator, is less than 1.**

## Improper Fractions

When the denominator of a fraction is less than its numerator, the fraction is called an improper fraction.

**Examples:**

$$\frac{3}{2}, \frac{5}{3}, \frac{17}{10}, \frac{9}{7}$$

*These fractions are improper fractions.*

The meanings of improper fractions and mixed fractions are similar. So we should talk about them together. But before the discussion, you should know how to represent a mixed fraction.

## Mixed Fractions

A mixed fraction is the combination of a whole number and a proper fraction.

$$2\frac{1}{2}$$

**Examples:**

$$2\frac{1}{2}, 1\frac{1}{3}, 4\frac{2}{7}, 11\frac{5}{9}$$

Whole number

Proper fraction

*These fractions are mixed fractions.*

In order to understand the meaning of the improper and the mixed fraction, we are mentioning the example of the chocolate bar once again. Imagine it is divided into 10 equal parts.

Since we divided the chocolate bar into 10 equal parts, we should write 10 as the denominator of the fraction representing the arbitrary number of parts, which you have considered.

We may consider an arbitrary improper fraction, which has a larger numerator than the denominator.

Say, the fraction is, $\dfrac{13}{10}$

The numerator represents the number of parts of the chocolate bar. Here, the numerator is 13, so the number represents 13 small parts of the chocolate bar.

But we know 10 such small parts make the whole chocolate. It means that 13 parts of chocolates correspond to 1 whole chocolate bar, along with 3 small parts of the chocolate.

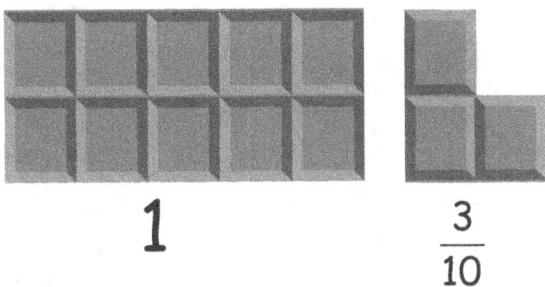

$$1 \qquad \frac{3}{10}$$

That 3 parts of chocolate can be represented by the fraction $^3/_{10}$.

And, 1 whole chocolate bar can be represented by 1, which is a whole number.

Therefore, the improper fraction, $^{13}/_{10}$ is a combination of a whole number, 1, and a proper fraction, $^3/_{10}$.

For the above reason, we may express the improper fraction as the following mixed fraction form.

$$1\frac{3}{10}$$

Alternatively, it can be said that a mixed fraction is the sum of a whole number and a proper fraction.

The present mixed fraction can also be represented as,

$$1\frac{3}{10} = 1 + \frac{3}{10}$$

**It is easy to convert an improper fraction into a mixed fraction, and vice versa. Some examples are given to show how it is done.**

$$\frac{7}{2}$$

It is an improper fraction because the numerator is greater than the denominator.

In order to make it a mixed fraction, we should divide the numerator by the denominator.

$$2\overline{\smash{)}7} \quad \begin{array}{r} 3 \\ \hline 7 \\ 6 \\ \hline 1 \end{array}$$

Here the quotient is 3, and the remainder is 1.

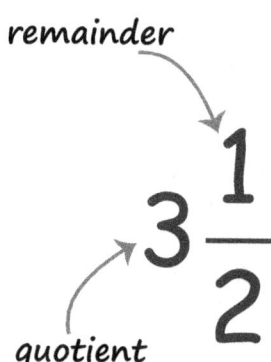

remainder

$$3\frac{1}{2}$$

quotient

Now you just put the obtained quotient in the place of the whole number, and the remainder as the numerator of the associated proper fraction. The denominator will remain the same.

**You may also do the reverse. A mixed fraction can be represented as an improper fraction too.**

$$3\frac{1}{2} \quad \longrightarrow \quad \frac{7}{2}$$

You need to multiply the whole number and the denominator, and add the numerator to the product to determine the corresponding improper fraction.

$$\text{Numerator} = 3\times2 + 1$$
$$= 7$$

$$\text{Denominator} = 2$$

Therefore, the corresponding improper fraction is

$$\frac{7}{2}$$

**You may try some more examples.**

| Improper Fraction | Mixed Fraction | Improper to Mixed | Mixed to Improper |
|---|---|---|---|
| $\frac{22}{7}$ | $3\frac{1}{7}$ | $7)\overline{\begin{array}{r}3\phantom{0}\\22\\21\\\hline1\end{array}}$ <br><br> Numerator: 1 <br> Denominator: 7 <br> Whole number: 3 | 3×7 + 1 = 22 <br><br><br><br> Numerator: 22 <br> Denominator: 7 |
| $\frac{7}{5}$ | $1\frac{2}{5}$ | $5)\overline{\begin{array}{r}1\phantom{0}\\7\\5\\\hline2\end{array}}$ <br><br> Numerator: 2 <br> Denominator: 5 <br> Whole number: 1 | 1×5 + 2 = 7 <br><br><br><br> Numerator: 7 <br> Denominator: 5 |
| $\frac{8}{3}$ | $2\frac{2}{3}$ | $3)\overline{\begin{array}{r}2\phantom{0}\\8\\6\\\hline2\end{array}}$ <br><br> Numerator: 2 <br> Denominator: 3 <br> Whole number: 2 | 2×3 + 2 = 8 <br><br><br><br> Numerator: 8 <br> Denominator: 3 |

Suppose your friend Bob brought three equal sized cakes to you. He just wants to play a mathematical game with you.

He cut the first cake into 5 identical pieces, the second cake into 7 identical pieces, and the third cake into 9 identical pieces. Then he gave you three different options, from which you were allowed to choose only one.

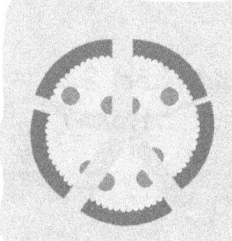 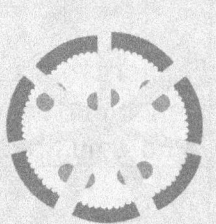 

**You may eat 2 pieces among the 5 from the first cake.**

Or,

**You may eat 3 pieces among the 7 from the second cake.**

Or,

**You may eat 4 pieces among the 9 from the third cake.**

**Which option will you choose in order to get the maximum amount of cake?**

You need to do a little mathematics to choose the appropriate option.

These three options are associated with three different fractions.

**If you selected the first option, you would get $^2/_5$ of the cake.**

**If you selected the second option, you would get $^3/_7$ of the cake.**

**And if you selected the third option, you would get $^4/_9$ of the cake.**

If you can tell the largest fraction among $^2/_5$, $^3/_7$, and $^4/_9$, you will eventually choose the desired option.

But you cannot compare the values of two or more fractions simply by looking at their numerators and denominators. You need to know a mathematical technique to compare them.

$$\frac{2}{5} \quad \frac{3}{7} \quad \frac{4}{9}$$

**In order to compare the values of fractions, first you need to express them with a common denominator. Then you compare their corresponding numerators.**
**But keep in mind that you should not alter the actual value of a fraction during the mathematical treatment.**

Let's learn how it is done.

The fractions of our interest are,

$$\frac{2}{5} \qquad \frac{3}{7} \qquad \frac{4}{9}$$

**Step 1**

Determine the lowest common multiple (L.C.M.) of the denominators.

Here the denominators are, 5, 7, and 9.

Their lowest common multiple is 315.

[If you do not know how to calculate the lowest common multiple (L.C.M.) of two or more numbers, learn from your math teacher or your parents.]

**Step 2**

Express these fractions with a common denominator.

Since 315 is the L.C.M. of the denominators, you are able to set this number as their common denominators.

You know that a fraction does not change its value, if we multiply an arbitrary number with both its numerator and denominator.

For example, consider a fraction $^2/_3$. If we multiply an arbitrary number, say 7, with its numerator and denominator, the value of the fraction does not change.

$$\frac{2}{3} = \frac{2 \times 7}{3 \times 7} = \frac{14}{21}$$

Both $^2/_3$ and $^{14}/_{21}$ have equal value. So they are called equivalent fractions.

We are going to apply the same technique to make the denominators 315 for the fractions $^2/_5$, $^3/_7$, and $^4/_9$. However, we have to determine the appropriate whole numbers, which will be multiplied to the numerators and the denominators keeping fractions' values unaltered.

First fraction,
$$\frac{2}{5}$$

We divide the L.C.M by its denominator to get the number, which has to be multiplied to its numerator and denominator.

L.C.M. ÷ denominator = 315 ÷ 5 = 63

Thus the fraction $^2/_5$ becomes,

$$\frac{2}{5} = \frac{2 \times 63}{5 \times 63} = \frac{126}{315}$$

Similarly, we can apply the same treatment for the second fraction, $^3/_7$.

Here the denominator is 7.

Thus,

L.C.M. ÷ denominator = 315 ÷ 7 = 45

The fraction, $^3/_7$ becomes,

$$\frac{3}{7} = \frac{3 \times 45}{7 \times 45} = \frac{135}{315}$$

The third fraction is $^4/_9$.

Thus,

L.C.M. ÷ denominator = 315 ÷ 9 = 35

The fraction, 4/9 becomes,

$$\frac{4}{9} = \frac{4 \times 35}{9 \times 35} = \frac{140}{315}$$

The alternative expressions of the fractions, $^2/_5$, $^3/_7$, and $^4/_9$ are $^{126}/_{315}$, $^{135}/_{315}$, and $^{140}/_{315}$ respectively.

$$\frac{126}{315} \qquad \frac{135}{315} \qquad \frac{140}{315}$$

Note that all the three fractions have a common denominator, 315.

## Step 3

Now you are allowed to compare the corresponding numerators as these fractions have equal denominators.

The original fractions were,

$$\frac{2}{5} \qquad \frac{3}{7} \qquad \frac{4}{9}$$

They can be rewritten with equal denominators as,

$$\frac{126}{315} \qquad \frac{135}{315} \qquad \frac{140}{315}$$

As their denominators are equal, we can compare their numerators, and arrange them in decreasing order.

$$\frac{140}{315} > \frac{135}{315} > \frac{126}{315}$$

Alternatively,

$$\frac{4}{9} > \frac{3}{7} > \frac{2}{5}$$

**Therefore you can conclude that the third option, which represents $^4/_9$ of a cake, will be the desired option for you.**

Can you arrange the following fraction in decreasing order?

$$\frac{1}{5}, \quad \frac{4}{15}, \quad \frac{2}{3}$$

**Solution:**

The L.C.M. of their denominators is 15.

Thus we can write these fractions setting their denominators to 15.

$$\frac{1}{5} = \frac{1 \times 3}{5 \times 3} = \frac{3}{15}$$

$$\frac{4}{15}$$ There is no need to alter this fraction as its denominator is already 15.

$$\frac{2}{3} = \frac{2 \times 5}{3 \times 5} = \frac{10}{15}$$

We can now arrange these fractions in decreasing order by comparing their numerators.

$$\frac{10}{15} > \frac{4}{15} > \frac{3}{15}$$

Or,

$$\frac{2}{3} > \frac{4}{15} > \frac{1}{5}$$

## Negative Numbers

So far in this book, we discussed whole numbers and fractions.

We may imagine a straight line, and visualize any number on it. This line is called the number line.

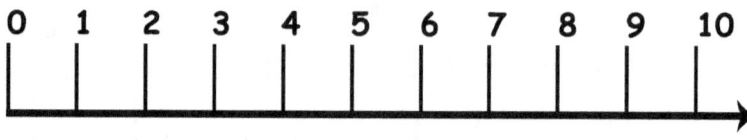

The above picture is quite similar to a ruler, which can be found in a geometry box. You usually use that ruler to measure the length of any object, or the distance between two arbitrary points.

The distance between any two consecutive numbers is the unit distance, and it should be measured in inches or centimeters.

If you carefully examine a ruler, you find it is marked with the numbers 0, 1, 2, 3, 4..., and each number is equally spaced to its adjacent numbers.

Suppose you are measuring the distance between points, A and B (as shown in the picture) with the ruler.

During the measurement, you place the 0 mark at the point A, and carefully point out the number where the point B is merged.

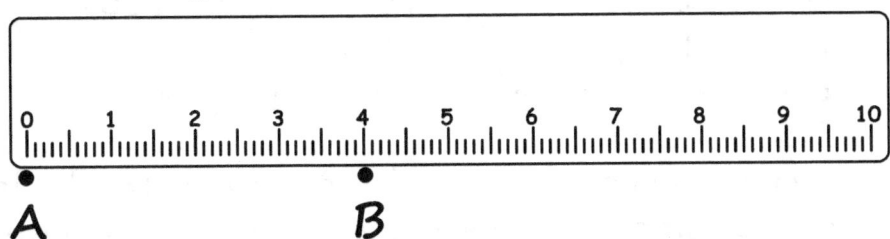

A

B

You see that point B has merged at the number 4 mark. Thus you may conclude the distance between A and B is 4 units (centimeters or inches).

Here you got a whole number, but it might be possible that point B lay in the middle of the marks 3 and 4.

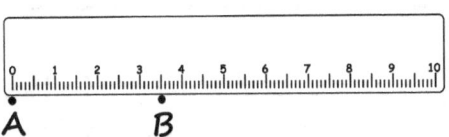

A

B

Then you would get a fractional number say 3.5 or 3½ units.

**The number line acts very similar to it.**

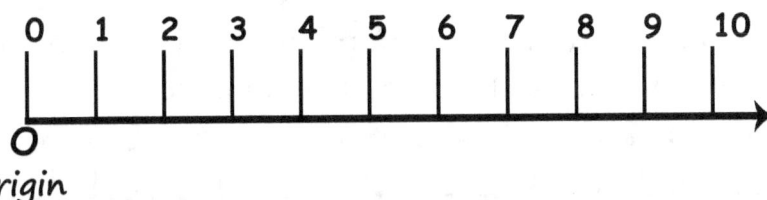

O

Origin

The number line starts from 0, so it is called the origin. As you move the right hand side from 0, you get the whole numbers 1, 2, 3, 4, 5, 6.... and so on.

Any position in between two consecutive whole numbers represents a fraction. That fraction can have any value depending on the position of an arbitrary point on the number line.

Therefore, if an arbitrary whole number or a fraction is given, you can easily tell its position on the number line.

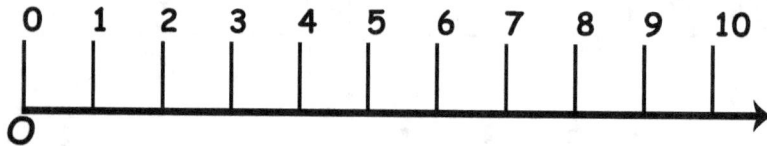

Every whole number or fraction lies at the right hand side of 0 on the number line. But why would we always move to the right hand side of 0 (origin)? What numbers will we get when we consider the left hand side of the origin?

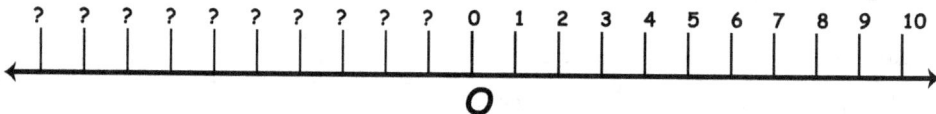

If you carefully examine the number line, you will find the numbers are gradually increasing as you move towards right, and when you move towards left, the numbers are gradually decreasing. Therefore, if you move to the left hand side of 0, you are going to get the numbers which are less than 0.

These numbers are negative numbers, and the complete number line will look like the following picture.

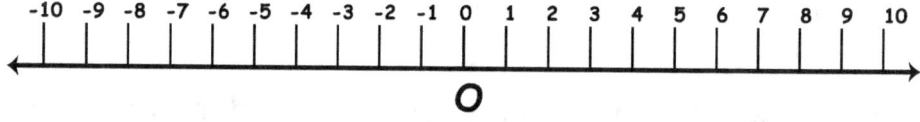

**A negative number is always represented with a negative sign written in front of it.**

-1 (Read as negative one or minus one)

-5 (Read as negative five or minus five)

-100 (Read as negative hundred or minus hundred)

On the other hand, the numbers at the right hand side of 0 are positive numbers. You may also use positive signs to express them.

+1 (Read as positive one or plus one)

+5 (Read as positive five or plus five)

+100 (Read as positive hundred or plus hundred)

Generally we do not show the positive sign when we write a positive number.
We generally write 1, instead of writing +1, or 5, instead of writing +5.

Therefore, when you see a number, which does not have any mathematical sign written before it, be confirm that it must be a positive number.

We already discussed how we are habituated with positive numbers in our daily life. We use those numbers to count objects or to measure any positive quantity.

On the other hand, negative numbers are not quite common in our everyday life. Yet in some cases, negative numbers are also important.

For example, the temperatures below zero degree are expressed by negative numbers.

We generally use the Celsius thermometer for measuring temperature. In this type of thermometer, the freezing and boiling point of water are considered the standard, and marked them as the numbers 0 and 100 respectively.

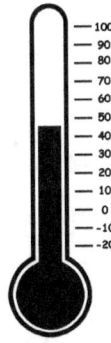

This temperature range is divided into 100 equal parts. Therefore, if we try to measure a certain temperature range, which is above the freezing point of water, we always get positive readings. However, sometimes we need to measure certain temperatures, which fall below the freezing point of water. As a result, we will get negative reading in a Celsius thermometer.

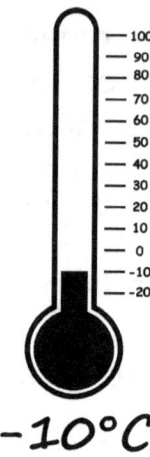

−**10°C**

It is true that the applications of negative numbers are very rare in our practical life, and these are used in a few cases, such as measuring very low temperatures. However, the concept of negative numbers is quite important in mathematics, especially in algebra. Now we are going to reveal the mathematical meaning of negative numbers.

**We mentioned earlier that negative numbers are never used to count things. Therefore, the following types of sentences are absurd in mathematics.**

He has -5 pen.

Bob eats -7 candy everyday.

**The quantities, such as mass, volume, length, etc. cannot be expressed by negative numbers.**
For example, the following sentences do not make any sense at all in mathematics.

There is -500 gram of meat in your bag.

Bob drinks -200 ml milk everyday.

Bob's house is -2 kilometers away from the school.

**On the other hand, the values of a few quantities like temperature, electric charges are sometimes expressed with negative numbers.**

> It is very cold outside. Today's temperature is -4 °C.

In the above statement, the temperature has been expressed with a negative number, -4, in the Celsius unit. Therefore, the value is telling us that the outside temperature is 4 degrees below from the freezing point of water.

There is no doubt that we cannot count physical objects, or measure non-negative quantities using negative numbers. However, we may write a statement in such a way where the negative number bears significant mathematical meaning, even when the negative number represents a number of objects or non-negative quantities.

Consider the following statement:

## Bob gave you -3 candies.

It is impossible to count the number of candies with a negative number. Therefore when you notice in this sentence that the number of candies has been denoted by a negative number, -3, you may think that the statement is incorrect.

In reality, the statement has a significant mathematical meaning. It is an obvious fact that Bob is not able to give you '-3 candies' because physical objects can not be counted using a negative number.

But if you write the statement in the following way, it becomes meaningful.

## You gave Bob 3 candies.

Now the statement is familiar to you, as it is possible to count 3 candies. Alternatively you can also say the incident as,

## You received –3 candies from Bob.

**'Someone gives you a negative number of something' actually means 'someone takes a positive number of something from you'.**

However, we do not generally talk like that in our practical life, but mathematically the statement is true.

**Some more examples can be shown.**

*He withdrew –1000$ from the bank yesterday.*

This statement actually means,

*He deposited 1000$ to the bank yesterday.*

*He moved –7 meters in the right direction.*

It means,

*He moved 7 meters in the left direction.*

You may construct many statements that way using your concept about negative numbers.

# EXERCISE 1

1) Arrange the following numbers in decreasing orders.

   (a) 234, 698, 5876, 42, 56, 627, 677.

   (b) 66787, 678899, 66729, 679919.

   (c) 7882, 62799, 89292, 349929, 72899.

   (d) 627, 7919, 4252, 9928, 4526.

   (e) 6838, 9982, 8288993, 7778828929, 8838388299.

2) Pick out all the odd numbers from below.

   56, 837, 87, 23, 34, 88, 26, 90, 57, 91, 928, 772, 2783,
   7829, 72, 9922, 827, 7711, 8288, 77289, 778685,
   45454, 562897, 798298, 8899, 8787, 754525,
   4765667.

3) Pick out all the prime numbers from below.

   55, 67, 2, 66, 67, 31, 59, 90, 88, 19, 45, 41, 37, 91, 27,
   5, 82, 78, 22, 7, 24, 23, 87, 29.

4) Pick out all the natural numbers from below.

   $-3$   6778   $\dfrac{1}{3}$   0   7   4.66

   23   $3\dfrac{1}{7}$   45   568788   $-345$

   4.666   67   89   $-45$   $5\dfrac{1}{5}$   18

5) Pick out all the rational numbers from below.

   $\dfrac{1}{4}$   3   3.22222222...   $5.\overline{7}$

   2.44948974...   4.000001   $3\dfrac{3}{4}$

   1.7320508075...   56776   $\pi$

6) Classify the following fractions as proper fractions, improper fractions, and mixed fractions.

$$3\frac{1}{7} \qquad \frac{23}{5} \qquad \frac{3}{25} \qquad 5\frac{4}{7}$$

$$6\frac{1}{3} \qquad \frac{7}{4} \qquad \frac{31}{7} \qquad \frac{34}{57}$$

$$\frac{7}{24} \qquad 8\frac{1}{3} \qquad \frac{19}{9} \qquad \frac{8}{9}$$

7) Convert the following improper fractions into mixed fractions.

$$\frac{13}{7} \qquad \frac{7}{2} \qquad \frac{22}{7} \qquad \frac{18}{5} \qquad \frac{67}{11} \qquad \frac{19}{18}$$

8) Convert the following mixed fractions into improper fractions.

$$2\frac{1}{7} \qquad 5\frac{4}{11} \qquad 3\frac{4}{7} \qquad 7\frac{1}{2} \qquad 9\frac{1}{3} \qquad 8\frac{5}{6}$$

9) Express the following fractions in their decimal forms.

$$\frac{1}{4} \qquad \frac{2}{3} \qquad 2\frac{3}{4} \qquad 3\frac{1}{5} \qquad \frac{3}{7} \qquad \frac{4}{5} \qquad \frac{5}{6}$$

10) Express the following decimals as the ratio of two whole numbers.

| | | |
|---|---|---|
| 0.07 | 1.333333... | 2.3 |
| 0.55 | 0.11 | $0.\overline{12}$ |
| 1.6 | 0.9 | 1.23 |
| $0.\overline{13}$ | 2.7 | 2.55555555.... |

11) Draw a number line, and show the positions of the following numbers on it.

$$-1.5 \qquad 7 \qquad 2\frac{1}{2} \qquad -4 \qquad -3.5$$

$$5 \qquad -6 \qquad -4\frac{1}{3} \qquad \frac{1}{2} \qquad -\frac{1}{2}$$

12) Arrange the following numbers in decreasing order.

(a) -5, 7, $-\frac{1}{3}$, $\frac{1}{3}$, 2.5, -7

(b) -3.99, -3.9, $-3\frac{1}{3}$, -3.909

(c) -30, 56, 0, -34, -0.001

(d) -7, -7.09, 7.001, $7\frac{1}{3}$, $-7\frac{2}{3}$

13) Consider the following statements, and tell which of them are mathematically meaningful. Write their actual meanings.

(a) He deposited -100 dollars in the bank.

(b) Alice has -6 dolls.

(c) Bob moved -90 meter towards south.

(d) Jack put -5 books in his bookshelf.

(e) Mary has -20 books.

(f) The tree had been -1 meter taller in the past year.

# CHAPTER 2  FOUR SIMPLE OPERATIONS:

## ADDITION, SUBTRACTION, MULTIPLICATION, AND DIVISION

You already learned in your school how to add, subtract, multiply, and divide numbers. In this chapter, we are going to discuss the same, but here you will learn some advanced topics of mathematics along with those four simple operations.

## What is addition?

**We generally add two or more numbers when we need to consider the total amount of some entities.**

Suppose you had some chocolates, and your friend Bob gave you some more on your birthday. How many chocolates do you have now?

You had

Bob gave

The simplest way to find the answer is as follows:

Put all the chocolates together, and count them all using the method, which you learned in the previous chapter.

Here you get 10 lines on the paper, thus you have 10 chocolates.

You may get the result using another technique without putting all the chocolates together.

First you counted you had 7 chocolates before.

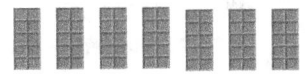

**You had 7 chocolates**

**Bob gave you 3 chocolates**

Then you counted the chocolates gifted by Bob. You got 3 chocolates from Bob.

Now you have to apply the concept of addition to calculate the total number of chocolates. If you manage to add 7 and 3, you will get the answer.

You draw 7 lines on the paper, and then you draw 3 more lines, which represent the chocolates gifted from Bob.

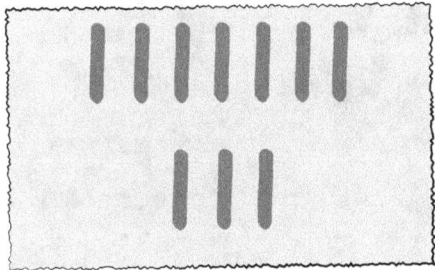

Again you find there are 10 lines on the paper.

The above mathematical operation is known as the addition, and this operation is represented by a '+' sign.

The operation of addition can be mathematically represented as,

$$7 + 3 = 10$$

Hope you have solved numerous addition-math problems in your school.

## What is subtraction?

**Subtraction has just the opposite meaning of addition. When you need to calculate the remaining number of objects after removal of some of them.**

We can demonstrate this operation with an example.

You have 10 chocolates, you decided you would eat 6 of them. How many chocolates will remain after the consumption?

You may answer the question following the practical method. That means you really ate 6 chocolates from them, and then counted the remaining chocolates in order to get the answer.

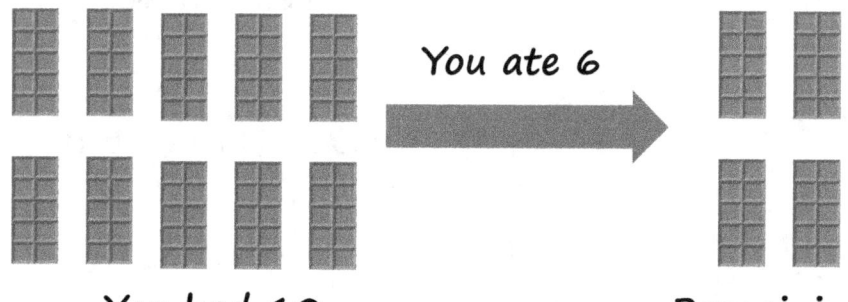

You ate 6

You had 10 chocolates

Remaining chocolates

Or you may apply a mathematical operation called subtraction to calculate the remaining chocolates.

First you draw 10 lines on a piece of paper to represent the initial number of chocolates.

Then strike out 6 lines among the 10 lines to indicate that you are going to eat 6 chocolates.

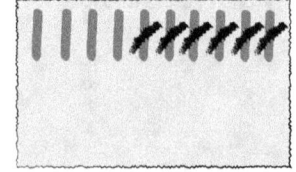

Now count the number of lines left on the paper. The number will be 4, you will conclude there will be 4 chocolates left after the consumption.

This mathematical treatment is termed as subtraction, and a negative sign '−' is used to express operation mathematically.

You may express the above mathematical treatment as,

$$10 - 6 = 4$$

## What is Multiplication?

We may depict a problem to understand the concept of multiplication.

Imagine you have 3 friends, Bob, Bill, and John. You want to give each of them 5 candies. How many candies are you going to buy from the store in order to distribute them?

You already know how to add numbers. Thus if you add the number of candies for all your friends, you definitely get the answer.

*Bob gets 5 candies.*
*Bill gets 5 candies.*
*John gets 5 candies.*

Therefore, the total number of candies will be the sum of these numbers.

$$5 + 5 + 5 = 15$$

Notice here, you added the same number several times.

In the present case, you added the number 5, three times. The meaning of such addition is the multiplication of 5 by 3.

$$5 \times 3 = 15$$

The above expression can also be written alternatively as,

$$3 + 3 + 3 + 3 + 3 = 15$$

Here you are adding the number 3 five times.

Thus, 5 + 5 + 5 and 3 + 3 + 3 + 3 + 3 have the same result, 15.

Any of the above two expressions is generally represented using a multiplication sign '×',

$$5 \times 3$$

Similarly, the meanings of following multiplications are given as,

**3 × 4 means**

> 4 + 4 + 4 (addition of 4 three times)
>
> or,
>
> 3 + 3 + 3 + 3 (addition of four 3)

**6 × 2 means**

> 6 + 6 (addition of two 6)
>
> or,
>
> 2 + 2 + 2 + 2 + 2 + 2
> (addition of six 2)

## Square, Cube, and Other Powers

You may multiply a given number with itself several times.

For example,

$$3 \times 3$$

(Here 3 is multiplied by 3.)

You can alternatively represent the above expression as,

$$3^2$$

The number at the superscript represents how many number of 3 are multiplied together.

$3^2$ means two 3 are multiplied together.

Similarly,

$3 \times 3 \times 3$ can be represented as $3^3$
$3 \times 3 \times 3 \times 3$ can be represented as $3^4$
$3 \times 3 \times 3 \times 3 \times 3$ can be represented as $3^5$

**The number written in the superscript is called 'power'.**

When the power is 2, it is also called square.

For example, $4^2$

You should read this expression as, 'square of four'

If the power is 3, the power is also called 'cube'.

$$2^3$$

You should read the above expression as, 'cube of two'.

When the power is greater than 3, you call it simply as, 'to the power'.

$5^6$     five to the power six

$4^8$     four to the power eight

## Square root, Cube root, and other roots

**The reverse operation of a 'power' is called 'root'.**

For example,
We know,
Square of 3 is 9.

$$3^2 = 9$$

Reverse operation
Square root of 9 is 3.

$$\sqrt{9} = 3$$

We generally use the following symbol to express the square root of a number.

√9 represents the square root of 9.

As you know that the square root of 9 is 3.

$$\sqrt{9} = 3$$

Similarly,

$$\sqrt{25} = 5 \qquad \sqrt{49} = 7$$
$$\sqrt{64} = 8 \qquad \sqrt{16} = 4$$

Likewise, cube roots and other roots are represented by the symbols.

$$\sqrt[3]{\phantom{x}} \qquad \sqrt[4]{\phantom{x}} \qquad \sqrt[6]{\phantom{x}}$$

and so on...

**Examples:**

$$2^3 = 8$$

We know the cube of 2 is 8. Therefore, the cube root of 8 is 2.

$$\sqrt[3]{8}$$

Similarly we know that 2 to the power 5 is 32.

$$2^5 = 32$$

Therefore, the fifth root of 32 is 2.

$$\sqrt[5]{32} = 2$$

Root of a number can also be represented as a power form. But in this case, the power should be expressed as the reciprocal of the corresponding root.

For example, the square root of 9 can be expressed as,

$$\sqrt{9} = 3 \quad \text{or,} \quad 9^{\frac{1}{2}} = 3$$

Both the expressions have the identical meaning.

Similarly,

$$8^{\frac{1}{3}} = 2 \qquad\qquad 27^{\frac{1}{3}} = 3$$

*Cube root of 8 is 2.*          *Cube root of 27 is 3.*

You will learn more about powers and roots in the indices chapter.

## What is division?

**You may consider division as the reverse operation of multiplication. To understand the concept, we should discuss the previous example in a different way.**

You bought 15 candies from the store. You equally distributed those candies among you three friends, Bob, Bill, and John. How many candies will each of them get?

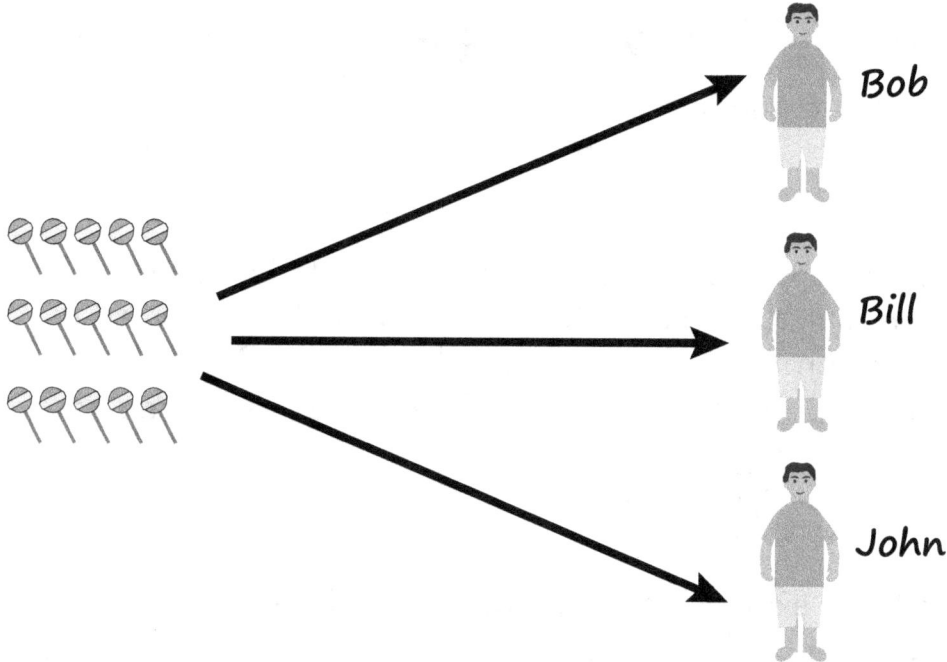

When we discussed multiplication, the number of obtained candies for each friend was given, and you were required to calculate the total number of candies.

This time, the total number of candies are given, and you have to calculate the number of candies for each friend. Thus you need to do just the reverse operation.

**When you divide the number 15 by 3, you get the answer, 5.**

$$15 \div 3 = 5$$

This operation is called the division, and any of the following symbols are used to represent this mathematical operation.

$$\div \qquad / \qquad -$$

Examples

$$15 \div 3 = 5$$
$$15 / 3 = 5$$
$$\frac{15}{3} = 5$$

You may practice more addition, subtraction, multiplication, and division problems from your school textbook.

- Addition operation shows commutative property. It means numbers can alter their positions when a + sign is written between them.
$$3 + 5 = 5 + 3$$
- Subtraction operation does not show commutative property. Numbers can not alter their positions when a - sign is written between them.
$$5 - 3 \neq 3 - 5$$
- Multiplication operation shows commutative property. Numbers can alter their positions when a × sign is written between them.
$$3 \times 5 = 5 \times 3$$
- Division operation does not show commutative property. Numbers can not alter their positions when a ÷ sign is written between them.
$$5 \div 3 \neq 3 \div 5$$

## Addition, Subtraction, Multiplication, and Division for Fractions

Now we should briefly discuss these four operations (addition, subtraction, multiplication, and division) in case of fractions. In various problems in algebra, we will apply these operations very soon.

## Addition and Subtraction of Fractions

We already discussed in the previous chapter how two or more fractions can be expressed with a common denominator. In order to add or subtract two fractions, first we need to express them with a common denominator, then we can add or subtract the numerators respectively to get the answer.

Let's understand the procedure with a few examples.

**Example:**

Evaluate the following sum

$$\frac{2}{5} + \frac{2}{7}$$

The L.C.M. of the denominators, 5 and 7, is 35.

Thus we can write the above expression in the following way.

$$= \frac{2 \times 7}{5 \times 7} + \frac{2 \times 5}{7 \times 5}$$

$$= \frac{14}{35} + \frac{10}{35}$$

Notice, now both their denominators become 35.

Now add their numerators to get the answer as a single fraction.

$$\frac{14}{35} + \frac{10}{35}$$
$$= \frac{14 + 10}{35}$$
$$= \frac{24}{35}$$

Therefore the sum of those two fractions is $^{24}/_{35}$.

Also, we can subtract two fractions expressing them with common denominators.

**Example:**

Evaluate the following.

$$\frac{2}{5} - \frac{2}{7}$$

Solution:

$$\frac{2}{5} - \frac{2}{7}$$
$$= \frac{2 \times 7}{5 \times 7} - \frac{2 \times 5}{7 \times 5}$$
$$= \frac{14}{35} - \frac{10}{35} \quad \text{(Expressing the fractions with equal denominators)}$$
$$= \frac{14 - 10}{35} \quad \text{(Subtracting their numerators)}$$
$$= \frac{4}{35}$$

A few more examples are shown.

**Example:**

Determine the following sum.

$$\frac{1}{2} + \frac{1}{4} + \frac{1}{8}$$

Solution:

L.C.M. (Lowest common multiple) of their denominators (2, 4, and 8) is 8.

Therefore,

$$\frac{1}{2} + \frac{1}{4} + \frac{1}{8}$$

$$= \frac{1 \times 4}{2 \times 4} + \frac{1 \times 2}{4 \times 2} + \frac{1}{8}$$

$$= \frac{4}{8} + \frac{2}{8} + \frac{1}{8}$$

$$= \frac{4 + 2 + 1}{8}$$

$$= \frac{7}{8} \qquad \text{(Answer)}$$

**Example:**

Solve the following subtraction:

$$\frac{1}{3} - \frac{1}{6}$$

Solution:

$$\frac{1}{3} - \frac{1}{6}$$

$$= \frac{1 \times 2}{3 \times 2} - \frac{1}{6} \qquad \text{(Since L.C.M. of 3 and 6 is 6.)}$$

$$= \frac{2}{6} - \frac{1}{6}$$

$$= \frac{2 - 1}{6}$$

$$= \frac{1}{6} \qquad \text{(Answer)}$$

64

**Example:**

Solve the following subtraction:

$$4 - \frac{1}{4}$$

Solution:

Here 4 is a whole number, so it can be expressed as a fraction form putting the numerator value 4, and the denominator value 1.

$$4 - \frac{1}{4}$$

$$= \frac{4}{1} - \frac{1}{4}$$

The L.C.M. of denominators, 1 and 4 is 4.

Therefore,

$$\frac{4}{1} - \frac{1}{4}$$

$$= \frac{4 \times 4}{1 \times 4} - \frac{1}{4}$$

$$= \frac{16}{4} - \frac{1}{4}$$

$$= \frac{16 - 1}{4}$$

$$= \frac{15}{4}$$

$$= 3\frac{3}{4} \quad \text{(Answer)}$$

## Multiplication and Division of Fractions

**Multiplication:** When we multiply two or more fractions, we get a new fraction, which is the product of them. In order to multiply fractions, you have to multiply their numerators, and write their product in the place of the numerator of the final product. And multiply their denominators, and write their product in the place of the denominator of the final product.

Let's multiply two fractions,

$$\frac{3}{5} \times \frac{2}{7}$$

$$= \frac{3 \times 2}{5 \times 7}$$

Product of numerators

Product of denominators

$$= \frac{6}{35} \quad \text{(Answer)}$$

**Example:**

Find the following product:

$$\frac{1}{2} \times \frac{2}{3} \times \frac{1}{4}$$

Solution:

$$\frac{1}{2} \times \frac{2}{3} \times \frac{1}{4}$$

$$= \frac{1 \times 2 \times 1}{2 \times 3 \times 4}$$

$$= \frac{2}{24} = \frac{1}{12} \quad \text{(Answer)}$$

**Division:** In order to divide a fraction by another fraction, you have to write the reciprocal of the divisor fraction changing the division sign into a multiplication sign. Their product will be the result of the division math problem.

To discuss the method, we consider the following example.

$$\frac{3}{4} \div \frac{2}{5}$$

Here dividend is $3/4$, and divisor is $2/5$.
Thus the reciprocal of divisor is $5/2$.

Dividend $\longrightarrow \dfrac{3}{4}$

Divisor $\longrightarrow \dfrac{2}{5}$

Reciprocal of divisor $\longrightarrow \dfrac{5}{2}$

Therefore,

$$\frac{3}{4} \div \frac{2}{5}$$
$$= \frac{3}{4} \times \frac{5}{2}$$
$$= \frac{15}{8} = 1\frac{7}{8} \quad \text{(Answer)}$$

We have written the reciprocal of divisor, and also changed the division sign into a multiplication sign.

**Example:** Solve the following division problem.

$$\frac{1}{2} \div \frac{1}{4}$$

Solution:

$$\frac{1}{2} \div \frac{1}{4}$$
$$= \frac{1}{2} \times \frac{4}{1} = \frac{1 \times 4}{2 \times 1} = \frac{4}{2} = 2 \quad \text{(Answer)}$$

**Example:** Solve the following division problem.

$$\frac{1}{5} \div 3$$

Solution:

$$\frac{1}{5} \div 3$$

$$= \frac{1}{5} \times \frac{1}{3} = \frac{1}{15} \quad \text{(Answer)}$$

**Example:** Solve the following division problem.

$$\frac{1}{3} \div \frac{5}{6}$$

Solution:

$$\frac{1}{3} \div \frac{5}{6}$$

$$= \frac{1}{3} \times \frac{6}{5}$$

$$= \frac{1 \times \cancel{6}^{2}}{\cancel{3} \times 5}$$

$$= \frac{2}{5} \quad \text{(Answer)}$$

# Addition and Subtraction for Negative Numbers

Now we should learn how to deal with these operations when negative numbers are associated. In order to understand this topic, we have to first discuss how numbers can be added or subtracted with the help of a number line.

Suppose we want to do the following calculation.

$$5 + 3$$

We know that the value of the above expression is 8, which we get by adding the numbers in the usual way. But this time, we want to solve it using the number line. Let's see how.

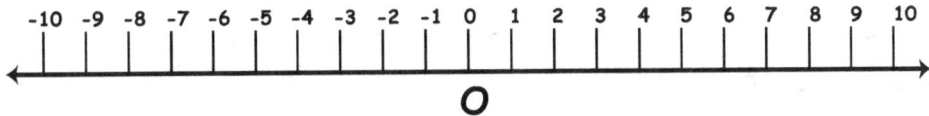

The first number is 5, so we should begin from 5 on the number line. Imagine you are standing at the 5 marked line. Now in order to add 3 to it, you should move 3 steps forward along the right hand side.

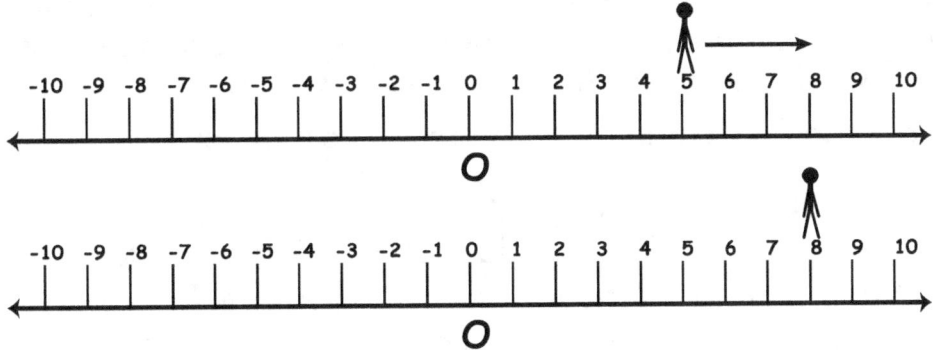

You stop at 8, thus the answer of this addition problem is 8.

We may also apply a similar technique to carry out subtraction problems. Consider the following subtraction,

## 5 - 3

In order to do the math, again you start from the position 5 on the number line. This time you should move 3 steps left because a negative sign is written in front of 3.

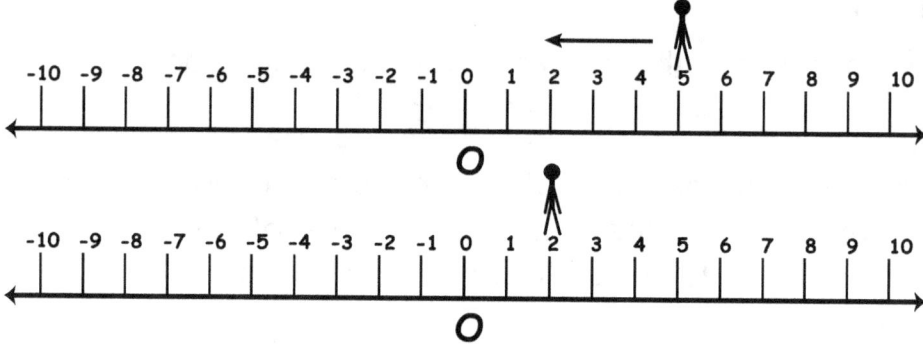

You stop at 2. Therefore, the answer to the above subtraction problem is 2.

If we carefully notice the number line, we find the value is increasing along the right direction, and the value is decreasing along the left direction.

**Therefore, it is quite expected when we are adding a number to a given number, we should move along right, as the total value increases after addition. On the other hand, when we subtract a number from a given number, we should move left to decrease the value.**

Now consider some more interesting examples.

## 5 - 7

In simple arithmetic, it is said that there is no way to subtract a larger number from a smaller one. But in algebra, this statement is not correct. In the above expression, we will see how 7 is subtracted from a smaller number 5 with the help of the number line.

# 5 - 7 = ?

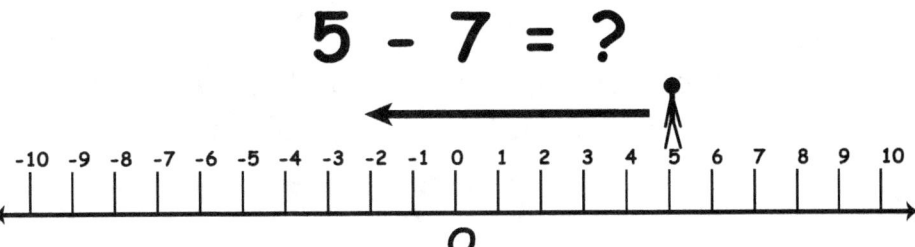

Here we want to subtract 7 from 5, so we should start from the position +5, and move 7 steps left on the number line.

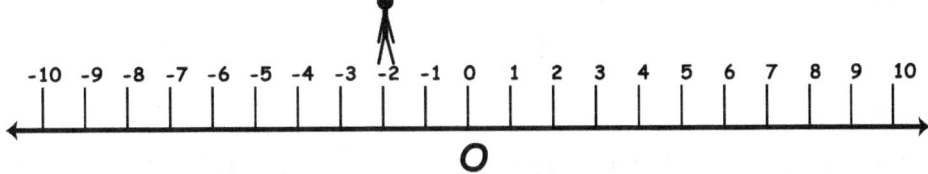

We stop at -2. The value of the above subtraction is -2.

# 5 - 7 = -2

You already know that -2 is a negative number. So when a larger number is subtracted from a smaller one, the result will be negative.

Find the value of the following expression with the help of the number line.

## -2 - 5

We should start from -2, and move 5 places left in order to get the result.

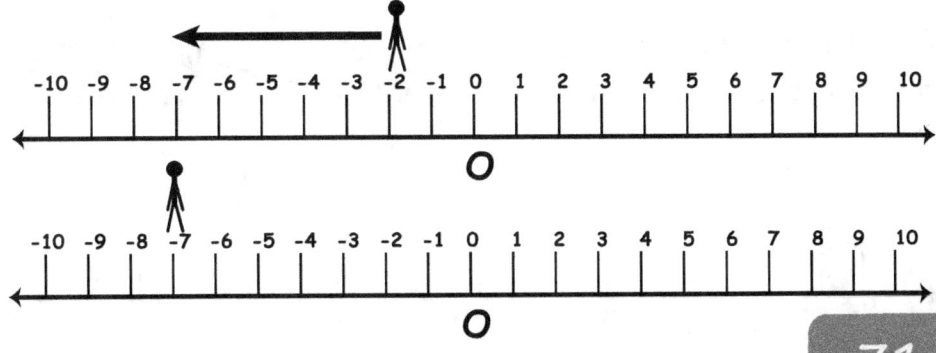

# -2 - 5

Finally we moved at the position -7 on the number line. Therefore, the value of the above expression is -7.

## -2 - 5 = -7

That way, we may solve larger addition-subtraction problems, which have more than one + or - signs.

Consider the following expression.

$$3 - 2 + 2 + 5 - 6$$

**Step 1**

We start from +3 and move 2 steps along the left direction because the second number is 2 with a negative sign. Thus our position becomes +1.

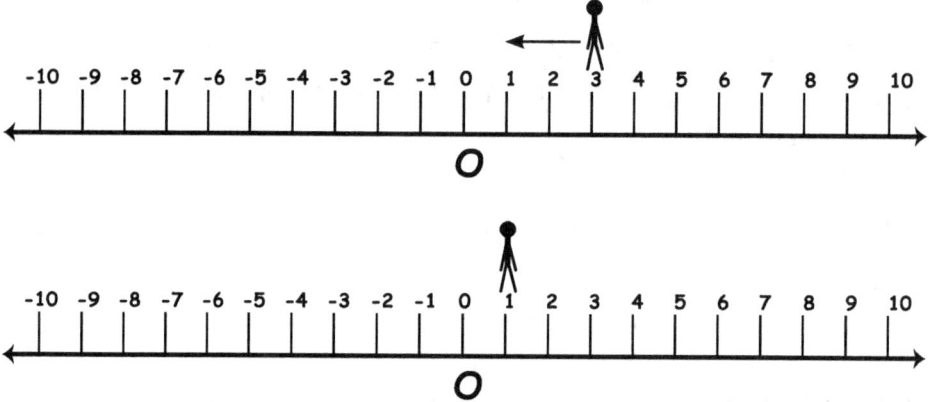

**Step 2**

Since the third number is 2 with a positive sign, we move 2 steps towards right from the current position +1. Now our position becomes +3.

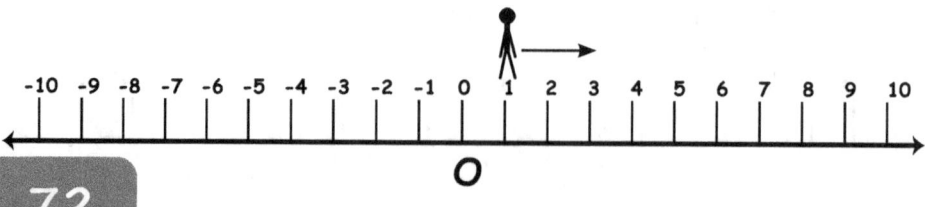

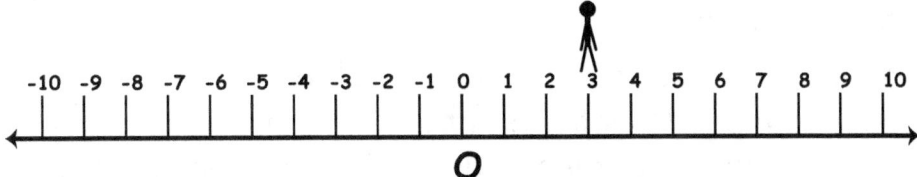

## Step 3

Then we move 5 steps right as the next number is 5 with a positive sign. Now our position changes to +8.

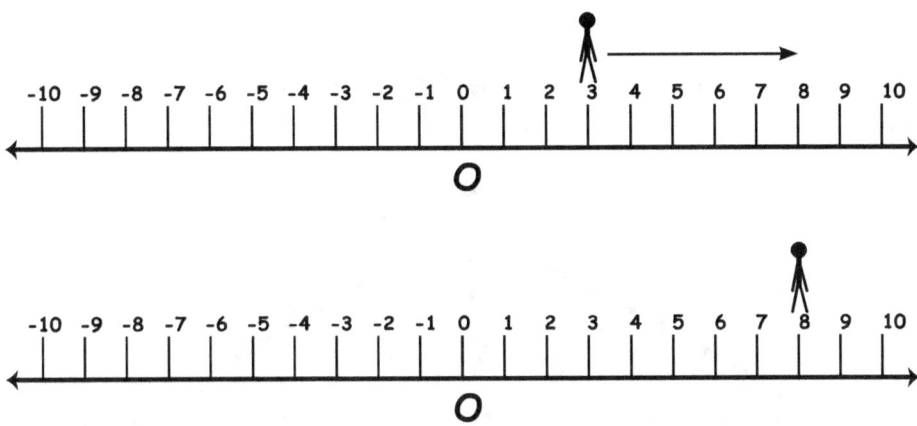

## Step 4

We should move 6 steps left as the last number is 6 with a negative sign, and our final position becomes +2.

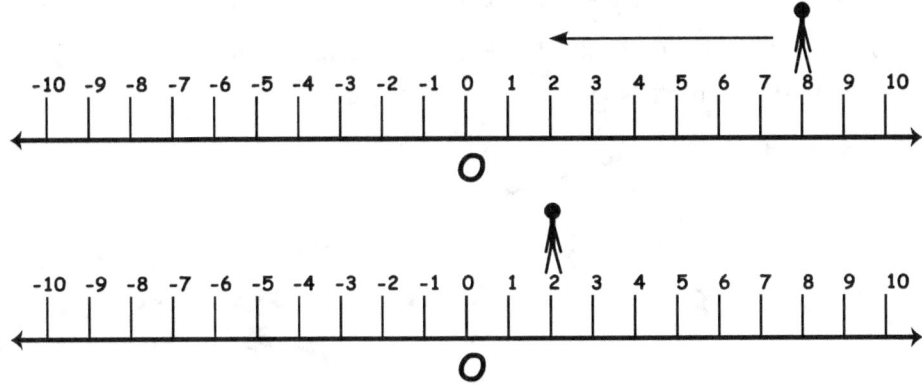

Therefore, the simplified value of the expression is 2.

$$3 - 2 + 2 + 5 - 6 = 2$$

## Some basic rules

It is possible to solve any addition-subtraction math problem with help of the number line. But when we encounter harder addition-subtraction problems, it will become time consuming. Therefore, we need to learn some simple rules in order to make the math problems easier.

● **We may alter the position of any number keeping its sign unchanged.**

For example, the given expression can be written in several ways.

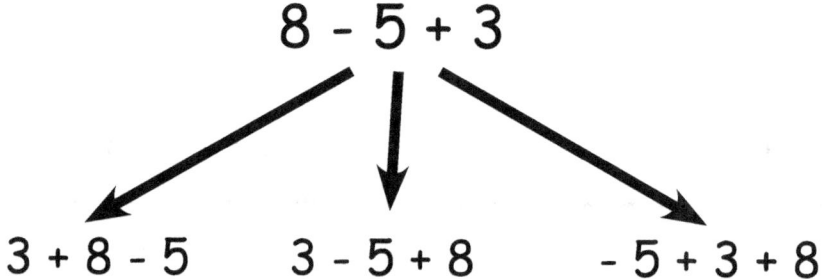

$$8 - 5 + 3$$

$$3 + 8 - 5 \qquad 3 - 5 + 8 \qquad -5 + 3 + 8$$

You should notice here that the sign written in front of each number has not been altered. The numbers have changed their positions along with their respective signs.

If you calculate each of the expressions using the number line, you will always get their value, 6.

$$3 + 8 - 5$$

$$3 - 5 + 8 \qquad\quad 6$$

$$-5 + 3 + 8$$

- **Use of parentheses is very efficient while we deal with negative numbers in a simplification math problem.**

We may put numbers inside parentheses, or take numbers out of parentheses while solving an addition-subtraction math problem. But we need to follow certain rules when we do such steps, because we need to carefully consider their signs.

**In case of positive sign written in front of parentheses**

Consider two numbers, one is associated with a positive sign, and other is associated with a negative sign.

$$+7 \text{ and } -5$$

When you write the numbers inside parentheses, and introduce a positive sign in front of the parentheses, you should write the expressions in the following way.

$$+(+7)$$

$$+(-5)$$

In this case, the numbers +7 and -5 did not change their signs.

Here,

$$+(+7) = 7$$

$$+(-5) = -5$$

**In case of negative sign written in front of parentheses**

When you write the numbers inside parentheses, and introduce a negative sign in front of the parentheses, you should write the expressions in the following way.

$$-(-7)$$

$$-(+5)$$

In this case, the numbers inside the parentheses changed their signs.

When you remove these parentheses, you get these numbers back with their initial signs.

$$-(-7) = 7$$

$$-(+5) = -5$$

**You probably noticed that when you introduce a negative sign before the parentheses, the signs associated with the numbers will alter. But when you write a positive sign before the parentheses, the numbers will not alter their signs.**

We will tell the reason later why the numbers change their signs inside parentheses associated with a negative sign.

Now we should understand the discussion through solving some problems.

Suppose we want to do the following calculation.

$$5 - 8$$

You already know how to calculate it using the number line. But this time, we will calculate the same without the help of the number line.

We can alter the positions of the numbers keeping their signs unchanged.

$$- 8 + 5$$

Now we write these numbers inside a bracket putting a negative sign before it.

$$- (8 - 5)$$

You should notice as we put a negative sign before the parentheses, the negative sign written before 8 is changed to positive, and the positive sign written before 5 is changed to negative.

We are now able to subtract 5 from 8 inside parentheses.

$$- (8 - 5)$$

$$= -(+3)$$

As we remove the parentheses, the final answer becomes,

$$-(+3)$$

$$= -3 \qquad \text{(Answer)}$$

Let's consider another example.

$$8 - 5 + 3 - 6 + 7 - 2$$

You might calculate the value using the number line, but it would be time consuming. We should apply the simplest technique to calculate this.

**Step 1**

We rearrange the numbers in the following way.

$$8 + 3 + 7 - 5 - 6 - 2$$

First we have written the numbers with positive signs, and then the numbers with negative signs.

Now we put these separate groups of numbers inside separate parentheses.

$$(8 + 3 + 7) - (5 + 6 + 2)$$

Notice that the numbers inside left hand side parentheses did not change their positive signs because a positive sign is considered in front of the bracket.

But the numbers written inside the right hand side parentheses have changed their signs from negative to positive because we put a negative sign before the parentheses.

The more elaborate representation of the expression is,

$$+ (+ 8 + 3 + 7) - (+ 5 + 6 + 2)$$

However, conventionally we do not show the positive sign of the leftmost number or parenthesis.

### Step 2

Now we can calculate the value of each parentheses.

$$(8 + 3 + 7) - (5 + 6 + 2)$$
$$= (18) - (13)$$
$$= 18 - 13$$
$$= 5$$

Therefore, the final answer of the given math problem is 5.

# Multiplication and Division for Negative Numbers

We previously discussed the meaning of multiplication for positive numbers. Now it is time to discuss how we perform multiplication when negative numbers are associated.

Suppose the following expression is given.

$$(+5)\times(-3)$$

Here a positive number, 5, is multiplied to a negative number, -3. Therefore, if we add -3 five times, we get the answer.

$$(-3) + (-3) + (-3) + (-3) + (-3)$$

Since, all the parentheses are associated with positive signs, the numbers inside the parentheses do not change their signs when we remove the parentheses.

$$- 3 - 3 - 3 - 3 - 3$$

Then we introduce another set of parentheses to make the numbers positive.

$$- (3 + 3 + 3 + 3 + 3)$$

*We should write a negative sign before the parentheses.*

$$= - (+15)$$
$$= - 15$$

Here we get the value - 15.

Therefore, $$(+5)\times(-3) = -15$$

Notice it is a negative number.

**When we multiply a negative number by a positive number, we get a negative number as their product.**

Now let's see what will be the answer when we multiply a negative number by another negative number.

$$(-5)×(-3)$$

We may understand this multiplication problem with the help of a pattern.

$$(+5)×(-3) = -15$$
$$(+4)×(-3) = -12$$
$$(+3)×(-3) = -9$$
$$(+2)×(-3) = -6$$
$$(+1)×(-3) = -3$$
$$(0)×(-3) = 0$$

We started with a number, +5, and decreased its value by 1 every time, and multiplied each of them by a negative number, -3. We notice the products follow an increasing order, and each time, their values are increased by 3. Therefore, if we continue to decrease the value of the multiplier by 1, the product will always increase by 3.

Therefore, the complete pattern will be,

$$(+5)×(-3) = -15$$
$$(+4)×(-3) = -12$$
$$(+3)×(-3) = -9$$
$$(+2)×(-3) = -6$$
$$(+1)×(-3) = -3$$
$$(0)×(-3) = 0$$
$$(-1)×(-3) = +3$$
$$(-2)×(-3) = +6$$
$$(-3)×(-3) = +9$$
$$(-4)×(-3) = +12$$
$$(-5)×(-3) = +15$$

Thus we obtain the desired result as,

$$(-5) \times (-3) = +15$$

**Therefore, multiplication of a negative number by another negative number gives a positive number.**

Now we should remember some simple rules of multiplication with respect to the signs of respective numbers.

(positive number)×(positive number) = positive product

(positive number)×(negative number) = negative product

(negative number)×(positive number) = negative product

(negative number)×(negative number) = positive product

Or simply we can write,

$$(+) \times (+) = (+)$$
$$(+) \times (-) = (-)$$
$$(-) \times (+) = (-)$$
$$(-) \times (-) = (+)$$

Some examples of multiplication examples are shown below:

$$(+3) \times (-2) = -6 \qquad (-3) \times (+8) = -24$$
$$(-4) \times (-4) = 16 \qquad (-3) \times (-6) = +18$$
$$(+5) \times (-4) = -20 \qquad (+5) \times (-6) = -30$$

You remember when we discussed addition and subtraction of negative numbers, we have shown that numbers change their signs when they are written inside a bracket when a negative sign is associated with the bracket.

But we did not tell the reason. Since you grew up with the concept of multiplication associated with negative numbers, now you are ready to understand that reason.

A negative sign written before a bracket is equivalent to the multiplication of the numbers inside the bracket by -1. So when you remove the bracket all the numbers inside the bracket are multiplied by -1. As a result they change their signs.

Example:

$$- (7 - 6) \text{ means } (-1) \times (7 - 6)$$

Division follows the same principle as multiplication. Hope you are able to understand it yourself.

(positive number)/(positive number) = positive quotient

(positive number)/(negative number) = negative quotient

(negative number)/(positive number) = negative quotient

(negative number)/(negative number) = positive quotient

Some division examples are shown below:

$$(-4) \div (+2) = -2$$
$$(15) \div (-5) = -3$$
$$(-14) \times (-7) = +2$$

# Simplification

As you have gone through the previous discussions, you are now able to solve numerous addition, subtraction, multiplication, and division math problems. But the discussions are not sufficient to solve the problems where mixtures of these mathematical operations are associated.

Suppose you and Bob, both were separately given the following math problem.

$$2 + 2 \times 2 - 2 \div 2$$

The given math problem is associated with addition, subtraction, multiplication, and division. Now Bob solved the problem in the following way.

$$2 + 2 \times 2 - 2 \div 2$$
$= 4 \times 2 - 2 \div 2$     (Bob first did addition.)
$= 4 \times 2 - 1$     (Then he did division.)
$= 8 - 1$     (Then he did multiplication.)
$= 7$     (Then he did subtraction.)

He got the answer, 7.

On the other hand you solved the problem in the following way.

$$2 + 2 \times 2 - 2 \div 2$$
$= 2 + 2 \times 2 - 1$     (First you did division.)
$= 2 + 4 - 1$     (Then you did multiplication.)
$= 6 - 1$     (Then you did addition.)
$= 5$     (Then you did subtraction.)

Thus you got the answer 5,

You got the answer 5, and your friend Bob got the answer 7 for the same math problem. Obviously, one of you solved the problem in the wrong way, otherwise there would not be different answers to such a math problem. For that reason, you have to follow some rules while simplifying complicated mathematical expressions.

## The BODMAS Rule

Now you realize that the mathematical operations, such as addition, subtraction, multiplication, and division, can not be performed in a random order. Therefore you have to maintain a definite order while dealing with these operations in a simplification problem.

BODMAS is the abbreviation of mathematical terms, and you have to follow the exact sequence as this abbreviation suggests, during a calculation.

B : Bracket
O : Order
D : Division
M : Multiplication
A : Addition
S : Subtraction

The first letter in BODMAS is B, and it is the short form of the word bracket. Thus when you are simplifying a mathematical expression, first you need to carry out the calculation inside a bracket or parentheses. Generally, brackets are represented by the following symbols.

( )        { }        [ ]

The second letter is O, and is the abbreviation of the word, 'order'. After carrying out the mathematical operations inside a bracket, you need to consider the order of powers of any number (such as $2^2$, $3^3$, $5^2$, etc.), and calculate their values.

Then comes D, M, A, S, which are abbreviations of division, multiplication, addition, and subtraction respectively. Therefore you should carry out the mathematical operations in the following order.

Division $\longrightarrow$ Multiplication $\longrightarrow$ Addition $\longrightarrow$ Subtraction

Now it is the time to deal with various simplification problems where we understand how the BODMAS rule actually works.

Let's continue our discussion with the previous mathematical expression.

$$2 + 2 \times 2 - 2 \div 2$$

Here we see brackets and power are absent. So we should skip B and O, and consider only the sequence D, M, A, and S.

We should carry out the division first.

$$2 + 2 \times 2 - 2 \div 2$$
$$= 2 + 2 \times 2 - 1$$

The next step will be multiplication.

$$2 + 2 \times 2 - 1$$
$$= 2 + 4 - 1$$

Then addition.

$$2 + 4 - 1$$
$$= 6 - 1$$

And at the end, we carry out subtraction.

$$6 - 1$$
$$= 5$$

Therefore, 5 is the simplified form of that mathematical expression. Therefore, we may conclude that you did the calculation in the right way, but Bob did not.

**Example:**
Simplify the following expression.

$$18 + (4 \div 2)^3 - 9$$

Solution:

$$18 + (4 \div 2)^3 - 9$$

| | |
|---|---|
| $= 18 + (2)^3 - 9$ | *(According to BODMAS rule, we first carried out the calculation inside the parenthesis.)* |
| $= 18 + 8 - 9$ | *(In the next step, we should calculate the value of $(2)^3$.)* |
| $= 26 - 9$ | *(Then addition)* |
| $= 17$ | *(At the end, we do subtraction.)* |

Therefore, the simplified value of that expression is 17.

**Example:**
Simplify the following expression.

$$(3 + 2) \times (8 - 3)$$

Solution:

$$(3 + 2) \times (8 - 3)$$

| | |
|---|---|
| $= (5) \times (5)$ | *(First we simplify the expression inside each bracket.)* |
| $= 25$ | *(Answer.)* |

**Example:**

Simplify the following expression.

$$4 + [8 + 4 \times \{8 \div (\frac{1}{4} \times 4)\}]$$

Solution:

$$4 + [8 + 4 \times \{8 \div (\frac{1}{4} \times 4)\}]$$
$$= 4 + [8 + 4 \times \{8 \div (1)\}]$$
$$= 4 + [8 + 4 \times \{8\}]$$
$$= 4 + [8 + 32]$$
$$= 4 + [40]$$
$$= 44 \qquad \text{(Answer.)}$$

# EXERCISE 2

1) Simplify the following expressions using the number line.

(a) 7 - 8

(b) 3 - 5 + 4

(c) 4 - 3 - 3 - 2 + 5

(d) 9 - 4 - 5

(e) - 7 - 2 + 6 - 1 + 5

(f) - 2 + 3 + 5 - 4 - 1

2) Simplify the following expressions.

(a) 18 - 21

(b) 20 - 33 + 5

(c) 14 - 4 + 5 - 11 + 15

(d) 900 - 400 - 500

(e) - 2.5 + 3.5 - 5.5

(f) - 44 + 3.55 + 11 - 4.55 - 19

3) Simplify the following expressions.

(a) (-3)×(-6)

(b) (-15)×(+3)

(c) (-3)×(-4)×(-2)

(d) (+2)×(-3)×(-2)×(+3)

(e) (-2.5)×(-4)

(f) (-2)×(-2)×(-2)×(-2)×(-2)

4) Simplify the following expressions.

(a) $\left(-\dfrac{2}{3}\right) \times \left(-\dfrac{3}{4}\right)$

(b) $3 - \dfrac{1}{2} - \dfrac{1}{4}$

(c) $2 \times \left(-\dfrac{3}{2}\right) + 2$

(d) $4 \div \left[\left(-\dfrac{1}{4}\right) \times \left(\dfrac{2}{3}\right)\right]$

(e) $4 - \left(\dfrac{1}{6} - \dfrac{1}{3}\right)$

5) Simplify the following expressions.

(a) $4 \div 2 \times (1 + 3)$

(b) $4 - [3 - 3 \times 3 + \{4 \times (5 - 9)\}]$

(c) $1 - [1 - \{1 - (1 - \sqrt{5-1})\}]$

(d) $\dfrac{4 - 2 \times 3 + 8}{1 + [4 - \{6 \div (5 - 3)\}]}$

(e) $\dfrac{3 \times 4 - 11}{5 \times 2 - 8} + \dfrac{4 \times 5 - 19}{3 \times 5 - 13}$

(f) $1 + \dfrac{1}{1 + \dfrac{1}{2}}$

(g) $\dfrac{1}{3} \div \dfrac{1}{3} + \dfrac{1}{3} \times \dfrac{1}{3} - \dfrac{1}{3}$

(h) $\dfrac{1}{1 + \dfrac{1}{1 + \dfrac{1}{2}}}$

(i) $(3 \div 4) \times 4 + (4 \div 5) \times 5 - (8 \div 9) \times 9$

(j) $\dfrac{3 \div 3 + 3 \times 3}{3 \times 3 - 3 \div 3}$

# CHAPTER 3 REPRESENTATION OF NUMBERS BY LETTERS

As you finished chapter 1 and 2, you are now able to carry out any type of mathematical calculations that are associated with numbers. The branch of mathematics that deals with numbers only, is known as arithmetic.
But this book is devoted to algebra, and you are going to learn this branch of mathematics this chapter onward.

We might have started this book straight from this chapter without discussing the previous chapters. If we did so, it would be difficult for you to understand the topic, because arithmetic and algebra follow the same mathematical principles, and it is impossible to learn algebra without knowing the basics of arithmetic.

In algebra, numbers are represented by letters (a, b, c, x, y, z, etc.), in place of actual numbers (1, 2, 3, 4, 5, etc.).
For example, in arithmetic, you are familiar with statements like these.

Bob eats 5 chocolates.

You have 18 books.

Bill's weight is 45.5 kg.

But when we are dealing with algebra, those statements may sound like the following.

Bob eats x number of chocolates.

You have y number of books.

Bill's weight is m kg.

But why would we use letters in place of numbers? Here we briefly answer the question. But as you continue reading this book, you will get deeper insight about the concept.

Imagine you have a cousin brother, who is 2 years younger than you. Now consider you do not know his age, but you know yours. If your age is 12 years, what is the age of your cousin?

You can easily determine the answer using your simple knowledge of arithmetic. Since he is 2 years younger than you, if you subtract 2 years from your age, his age can be determined.

$$12 - 2 = 10$$

Therefore, your cousin is 10 years old.

That way you can determine your cousin's age in any past or future year if you know your own age at that year.

As you notice your age and your cousin's age are not fixed numbers, but they vary with years, so these are called variables. **If a quantity varies with respect to another quantity, the quantity is called a variable.** Since both of your ages are varying every year, they can be considered as variables. However, one condition will always be true in the case of your ages that your age is always 2 years greater than your cousin's age.

Now you may construct a simple equation representing your age and your cousin's age with letters instead of numbers.

Let your age is x years, and your cousin's age is y years. Therefore,

$$y = x - 2$$

This equation tells you that your cousin's age, y is 2 years less than your age, x.

With the help of the above equation, you may answer the following questions.

> What was your cousin's age when you were 5 years old?

You have to just put 5 in the place of x in order to find y.

$$y = 5 - 2$$

or, y = 3

Thus your cousin was 3 years old.

> What will be your cousin's age when your age will reach 15 years?

Following the same method, you can answer.

$$y = 15 - 2$$

or, y = 13

He will be 13 years old.

The above example was quite easy because you can apply the knowledge of arithmetic also in order to answer those questions. However, there are many problems in mathematics where you need to exclusively apply algebra to solve such problems.

We will soon return to this topic in our upcoming chapters.

But first of all you have to learn how to deal with algebraic expressions when different mathematical operations, such as addition, subtraction, multiplication, and division, are carried out.

## Addition and Subtraction in Algebra

Suppose a glass jar contains x number of candies, and another glass jar contains y number of candies. You poured all these candies into a third jar. How many candies are there in the third jar?

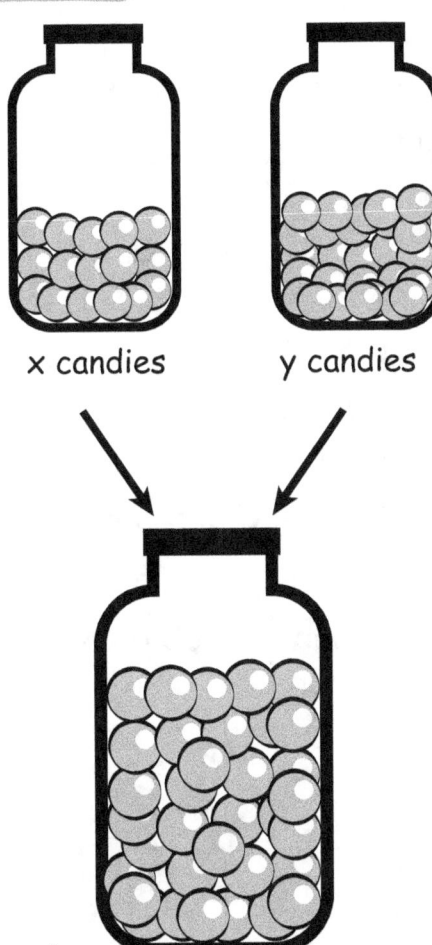

x candies        y candies

If x and y were represented by numbers, you might add them up to get the answer. But here the number of candies are represented by the letters, x and y. Therefore, you should represent their sum as the following expression.

### x + y

Since you do not know the actual value of x and y in numbers, you may conclude that the third jar contains (x + y) number of candies.

(x + y) candies

Similarly, if you want to subtract b from a, you should write the following expression.

$$a - b$$

A few more examples are shown below.

**Example:**
Express the sum of p, q, and r in algebra.

Solution:

$$p + q + r$$

**Example:**
Statement: c is subtracted from the sum of a and b.
Solution:

Sum of a and b is represented as,

$$a + b$$

When c is subtracted from their sum, the expression becomes,

$$(a + b) - c$$

$$= a + b - c$$

**Example:**
Statement: Sum of x and y is subtracted from the sum of w and z.

Solution:

Sum of x and y is,

$$x + y$$

And sum of w and z is,

**w + z**

When the sum of x and y is subtracted from the sum of w and z, the expression becomes,

$$(w + z) - (x + y)$$

Notice that here we put the sums inside brackets, otherwise there could have possibility of writing incorrect signs.

If the brackets are removed, the expression becomes,

$$(w + z) - (x + y)$$
$$= w + z - x - y$$

Note that both x and y altered their signs as we removed the bracket. We already discussed the reason in the previous chapter.

## Multiplication and Division in Algebra

The multiplications of two or more number in arithmetic are generally expressed as,

**3×5**     (Here 3 is multiplied by 5.)

**2×7×8**     ( Here 2 is multiplied by 7 and 8.)

In arithmetic, the multiplication sign is represented by the × symbol. But the multiplication sign in algebra is quite different. It is represented by a dot (.) symbol. Even in some cases, multiplication does not require any symbol to represent. If you write two letters, one beside of another without any gap, it also represents multiplication.

For example, the multiplication of a and b can be represented in the following way.

$$a.b \qquad or \qquad ab$$

We may also represent multiplication of a letter by a number.

Multiplication of 3 and a.

$$3.a$$

Or you may also write as,

$$3a$$

However, you should not write the above expression as a3, because a number is always written on the left side by convention.

Some expressions of multiplied forms in algebra are shown below.

| Arithmetic form | Algebra form |
|:---:|:---:|
| a×b×c | abc |
| a×3×5×c | 15ac |
| 8×4×x | 32x |

Also we generally do not use ÷ sign in order to represent division in algebra. Instead, a slash (/) or a horizontal line is used here to represent division.

For example,

p is divided by q.

The above statement is represented in algebra as,

$$p/q \qquad or \qquad \frac{p}{q}$$

We already used these division signs to represent fractions in arithmetic.

## Square, Cube, and Other Powers

Previously we discussed the meaning of square, cube, and power of any number in mathematics.

For example, the meaning of $2^4$ is four 2 are multiplied together.

$$2^4 = 2×2×2×2 = 16$$

The same expression is also applicable in algebra. For example, when we write $a^4$, it means four a are multiplied together.

$$a^4 = a.a.a.a$$

$x^5$ means, five x are multiplied together.

$$x^5 = x.x.x.x.x$$

Now we are going to solve some mathematical problems in order to understand how to deal with algebraic calculations.

**Example:**
Simplify the following sum.

$$3a + 2a$$

Solution:
The first term of the above expression is 3a. It means a is multiplied by 3. Alternatively, we can say that 3a is the sum of three a.

$$3a = a + a + a$$

Similarly, we can express the second term 2a as,

$$2a = a + a$$

Now the expression 3a + 2a becomes,

$$3a + 2a = a + a + a + a + a$$

$$\underbrace{\qquad\qquad}_{3a} \underbrace{\qquad}_{2a}$$

The expression is the sum of five a, which is nothing but five times a.

Therefore,

$$3a + 2a = a + a + a + a + a$$
$$= 5a$$

$$\boxed{3a + 2a = 5a}$$

**Here 3, 2, and 5 are called coefficients of a. When we encounter such situations in algebra, we can add or subtract coefficients of like terms to simplify an expression.**

**Example:**
Simplify the following expression.

$$8x + 5x - 3x - 2x$$

Solution:

We can introduce parentheses to the expression, and rewrite as,

$$8x + 5x - 3x - 2x$$
$$= (8x + 5x) - (3x + 2x)$$

*The third and the fourth term changed their signs inside the bracket because the introduced bracket is associated with a negative sign.*

$$(8x + 5x) - (3x + 2x)$$

$$= (13x) - (5x)$$

$$= 8x \quad \text{(Answer)}$$

**Example:**
Express the following multiplication in a simplified form.

$$3 \times a \times 2 \times a \times b \times a$$

Solution:

Alternatively we may write the expression as,

$$3 \times a \times 2 \times a \times b \times a$$

$$= 3 \times 2 \times a \times a \times a \times b \quad \text{(We just altered their order.)}$$

$$= 6.a.a.a.b$$

$$= 6a^3b$$

The simplified form is $6a^3b$.

**Example:**
Simplify the following expression.

$$(2a^2 + 3a) + (a^2 + a) + (3a^2 - 2a) - (a^2 + 2a)$$

Solution:

$$(2a^2 + 3a) + (a^2 + a) + (3a^2 - 2a) - (a^2 + 2a)$$

$$= 2a^2 + 3a + a^2 + a + 3a^2 - 2a - a^2 - 2a \quad \text{[Removing brackets]}$$

$$= 2a^2 + a^2 + 3a^2 - a^2 + 3a + a - 2a - 2a \quad \text{[Rearranging like terms]}$$

$$= 5a^2 + 0.a = 5a^2 \quad \text{[Answer]}$$

**Example:**

Multiply ax and bx.

Solution:

$$(ax) \times (bx)$$

$$= a.x.b.x$$

$$= a.b.x.x$$

$$= abx^2 \qquad \text{[Answer]}$$

**Example:**

Multiply 3xy and 2y.

Solution:

$$(3xy) \times (2y)$$

$$= 3.x.y.2.y$$

$$= 3.2.x.y.y$$

$$= 6xy^2 \qquad \text{[Answer]}$$

**Example:**

Mathematically represent the following statement.

Sum of x and y is divided by z.

Solution:

Sum of x and y is represented as,

$$x + y$$

We should divide the sum by z. Therefore the expression becomes,

$$\frac{x + y}{z}$$

**Example:**

Given $a = 1$, $b = 2$, $c = 3$. Find the value of

$$\frac{ab + c}{b} + \frac{a + c^2}{b^2}$$

Solution:

$$\frac{ab + c}{b} + \frac{a + c^2}{b^2}$$

Put the values of $a$, $b$, and $c$ in the above expression, then simplify as usual.

$$\frac{ab + c}{b} + \frac{a + c^2}{b^2}$$

$$= \frac{1 \times 2 + 3}{2} + \frac{1 + 3^2}{2^2}$$

$$= \frac{2 + 3}{2} + \frac{1 + 9}{4}$$

$$= \frac{5}{2} + \frac{10}{4}$$

$$= \frac{5}{2} + \frac{5}{2}$$

$$= \frac{5 + 5}{2}$$

$$= \frac{10}{2} = 5 \text{ [Answer]}$$

# EXERCISE 3

1)  Mathematically express the following statements:

(a) Sum of p and q.

(b) Subtract x from w.

(c) Product of a and x.

(d) Divide x by y.

(e) Subtract c from the sum of a and b.

(f) Product of a, b, and c.

(g) Divide the product of a and c by b.

(h) Divide the sum of x and y by the sum of a and b.

(i) Product of ab and ac.

2)  Write the algebraic forms of the following expressions:

| | | | |
|---|---|---|---|
| (a) | $a \times b$ | (k) | $(2 \times a + 3 \times b) \times (a + b)$ |
| (b) | $m \div n$ | (l) | $a \times a + 2 \times a$ |
| (c) | $p \times q \times r$ | (m) | $(a \div b) + (b \div a)$ |
| (d) | $(a \times b) \div c$ | (n) | $x + (1 \div x)$ |
| (e) | $a \times (b \div c)$ | (o) | $\{a \times (a + b)\} \div (x + y)$ |
| (f) | $(a + b) \div (x + y)$ | (p) | $3 \times a \times a \times a \times b + 2 \times a \times a \times b \times b$ |
| (g) | $2 \times a \times b \times 5 \times b$ | (q) | $(a + b) \times (a + b) \times (a + b)$ |
| (h) | $2 \times a + 3 \times b$ | (r) | $a \times a + 2 \times a \times b + b \times b$ |
| (i) | $a \times (b + c)$ | (s) | $a \div (b + c + d)$ |
| (j) | $a \times a \times a \times a \times a \times a$ | (t) | $(2 \times a + 3 \times b + 5 \times a \times b) \times b \times b$ |

3) Simplify the following expressions:

(a)   $2a \cdot 3a$

(b)   $a + 2a$

(c)   $3a^2 \cdot 5b^2$

(d)   $2x + 3x - 7x + 4x$

(e)   $2a \cdot 2b$

(f)   $(a + 2b) + (3a - 3b) - (5a + 4b)$

(g)   $2x - [3x - \{2x - (3x - x)\}]$

(h)   $(5p^2 + 2p) - (p^2 - 3p) + (6p^2 - p)$

(i)   $6a^2b^3 \cdot \dfrac{1}{3ab}$

(j)   $(a^2 + 5ab + 2b^2) + (2a^2 + ab + 3b^2)$

(k)   $(a^2 + 2a + 1) + (2a^2 + 3a + 2) - (a^2 + 2a + 4)$

(l)   $\dfrac{3a^2b}{2c^2} \cdot \dfrac{4c^3b}{3a} \cdot \dfrac{3ac}{b^3}$

(m)   $\dfrac{x + y}{x + y + z} + \dfrac{y + z}{x + y + z} + \dfrac{z + x}{x + y + z}$

4) Given $x = 2$,
Determine the value of the following expression.

$$x^3 + 2x^2 + 3x + 2$$

5) Given $a = 1, b = 2, c = 3$
Determine the value of the following expression.

$$\frac{a}{b} - \left( \frac{a + b}{c} - \frac{a + c}{bc} \right)$$

# CHAPTER 4  SOME USEFUL RELATIONS AND FORMULAS

Since you completed chapter 3, you now know how to deal with addition, subtraction, multiplication, and division in algebra. However, in order to carry out harder problems in algebra, you have to learn a few more mathematical techniques. In this chapter, we are going to talk about them.

## Relation 1

$$x(a + b) = xa + xb$$

The expression in the left hand side of this equation is the product of x and (a + b).
[(a + b) represents the sum of a and b.]
And the right hand side of the equation is the sum of two products, xa and xb.

We can apply this relation in many problems in algebra. Soon we are going to solve some mathematical problems, but first of all we should check the validity of that equation.

We may consider x, a, and b as three arbitrary positive integers, say 3, 2, and 4 respectively.

$$x = 3$$
$$a = 2$$
$$b = 4$$

Thus left hand side of the equation can be expressed as,

$$3 \times (2 + 4)$$

And right hand side of the equation can be expressed as,

$$3 \times 2 + 3 \times 4$$

The simplified forms of the right hand side and the left hand side are found to be equal when we determine their values using the BODMAS rule.

| Left hand side | Right hand side |
|---|---|
| $3 \times (2 + 4)$ | $3 \times 2 + 3 \times 4$ |
| $= 3 \times (6)$ | $= 6 + 12$ |
| $= 18$ | $= 18$ |

Therefore,

Left hand side = Right hand side

If you considered any other values of x, a, and b, you would get the same result as well. We may prove the validity of the relation in another way.

x(a+b) means,

$(a + b) + (a + b) + (a + b) + (a + b) + (a + b) + \ldots$ x times

$= (a + a + a + a + a + \ldots \text{x times}) + (b + b + b + b + b + \ldots \text{x times})$

$= xa + xb$    [Proved]

A similar relation will be also valid if the positive signs are replaced by negative signs.

$$x(a - b) = xa - xb$$

You may check its validity applying the similar method.

Now we should solve some simple mathematical problems.

**Example:**

Simplify:

$$\frac{3x + 6}{x + 2}$$

Solution:

The numerator of the expression can be written as,

$$3x + 6$$

$$= 3x + 3 \times 2$$

$$= 3(x + 2) \qquad \text{[Applying the relation } x(a + b) = xa + xb]$$

Therefore,

$$\frac{3x + 6}{x + 2}$$

$$= \frac{3(x + 2)}{x + 2} \qquad \text{[Here we considered, } (x + 2) \neq 0]$$

$$= 3 \qquad \text{[Answer]}$$

**Example:**

Simplify:

$$\frac{4 - 2x}{x - 2}$$

Solution:

$$\frac{4 - 2x}{x - 2}$$

$$= \frac{2(2 - x)}{x - 2} \qquad \text{Notice how we altered the signs here.}$$

$$= \frac{-2(x - 2)}{x - 2} \qquad \text{[Considering, } (x - 2) \neq 0]$$

$$= -2 \qquad \text{[Answer]}$$

**Example:**

Simplify:

$$x(x + a) - x^2$$

Solution:

$$x(x + a) - x^2$$
$$= x^2 + ax - x^2$$
$$= ax \qquad \text{[Answer]}$$

**Example:**

Simplify:

$$a(b - c) + b(c - a) + c(a - b)$$

Solution:

$$a(b - c) + b(c - a) + c(a - b)$$
$$= ab - ac + bc - ab + ac - bc$$
$$= ab - ab + bc - bc + ac - ac$$
$$= 0 \quad \text{[Answer]}$$

## Relation 2

$$(a + b)(x + y) = ax + bx + ay + by$$

It is easy to prove the above relation with help of the previous relation. Let's see how.

$$(a + b)(x + y)$$

Let,

$$(a + b) = p$$

Thus the expression becomes,

$$(a + b)(x + y)$$

$$= p(x + y)$$

The above expression can be expanded using the previous relation as,

$$p(x + y)$$

$$= px + py$$

Putting the value of p in the above expression we get,

$$px + py$$

$$= x(a + b) + y(a + b) \qquad [\text{ Since } p = a + b]$$

Expanding the relation once again, we get,

$$x(a + b) + y(a + b)$$

$$= ax + bx + ay + by \qquad [\text{Proved}]$$

Now we are able to solve several multiplication problems applying this relation.

**Example:**

Multiply:

$$(x + 1)(x + 2)$$

Solution:

$$(x + 1)(x + 2)$$
$$= x(x + 1) + 2(x + 1)$$
$$= x^2 + x + 2x + 2$$
$$= x^2 + 3x + 2 \qquad \text{[Answer]}$$

**Example:**

Multiply:

$$(x + 2)(x - 3)$$

Solution:

$$(x + 2)(x - 3)$$
$$= x(x + 2) - 3(x + 2)$$
$$= x^2 + 2x - 3x - 6$$
$$= x^2 - x - 6 \qquad \text{[Answer]}$$

At this point we are ready to learn some important formulas which are widely used in algebra.

**Formula 1**

$$(a + b)^2 = a^2 + 2ab + b^2$$

**Formula 2**

$$(a - b)^2 = a^2 - 2ab + b^2$$

**Formula 3**

$$a^2 + b^2 = (a + b)^2 - 2ab$$
$$= (a - b)^2 + 2ab$$

**Formula 4**

$$a^2 - b^2 = (a + b)(a - b)$$

**Formula 5**

$$4ab = (a + b)^2 - (a - b)^2$$

We are now going to prove these formulas one by one using our basic knowledge of algebra.

## Proof of formula 1

$$(a + b)^2$$

The expression can be written in the product form.

$$(a + b)^2$$
$$= (a + b)(a + b)$$
$$= a(a + b) + b(a + b)$$
$$= a^2 + ab + ab + b^2$$
$$= a^2 + 2ab + b^2$$

Therefore the relation has been proved.

$$(a + b)^2 = a^2 + 2ab + b^2$$

## Proof of formula 2

The proof of formula 2 is quite similar to the previous one.

$$(a - b)^2$$

The expression can be written in the following product form.

$$(a - b)^2$$
$$= (a - b)(a - b)$$
$$= a(a - b) + b(a - b)$$
$$= a^2 - ab - ab + b^2$$
$$= a^2 - 2ab + b^2 \qquad \text{[Proved]}$$

## Proof of formula 3

$$a^2 + b^2$$

The above expression can be written as,

$$a^2 + b^2$$

$$= a^2 + b^2 + 2ab - 2ab$$

Note that here we introduced two extra terms, $+2ab$ and $-2ab$. They do not alter the overall value of the expression because total value of these two terms is 0, and if we add or subtract 0, the value does not alter for any expression.

$+2ab - 2ab = 0$

Therefore,

$$a^2 + b^2 + 2ab - 2ab$$

$$= (a^2 + b^2 + 2ab) - 2ab$$

[ We know, $a^2 + b^2 + 2ab = (a + b)^2$]

$$= (a + b)^2 - 2ab$$

Also,

$$a^2 + b^2 + 2ab - 2ab$$

$$= a^2 + b^2 - 2ab + 2ab$$

$$= (a^2 + b^2 - 2ab) + 2ab$$

[ We know, $a^2 + b^2 - 2ab = (a - b)^2$]

$$= (a - b)^2 + 2ab$$

Therefore, we obtained following two mathematical relations.

$$a^2 + b^2 = (a + b)^2 - 2ab$$

$$a^2 + b^2 = (a - b)^2 + 2ab$$

## Proof of formula 4

We have to prove,

$$a^2 - b^2 = (a + b)(a - b)$$

It will be easier to prove the equation if we expand the right hand side expression, and show it finally gives the expression at left hand side.

*Right hand side*

$$(a + b)(a - b)$$

$$= a(a - b) + b(a - b)$$

$$= a^2 - \cancel{ab} + \cancel{ab} - b^2$$

$$= a^2 - b^2 \qquad \text{[It is the expression of Left hand side.]}$$

Since,

$$L. H. S = R. H. S.$$

The following mathematical expression has been proved.

$$a^2 - b^2 = (a + b)(a - b)$$


## Proof of formula 5

We have to prove the following,

$$4ab = (a + b)^2 - (a - b)^2$$

*Right hand side*

$$(a + b)^2 - (a - b)^2$$

$$= (a^2 + 2ab + b^2) - (a^2 - 2ab + b^2)$$

$$= \cancel{a^2} + 2ab + \cancel{b^2} - \cancel{a^2} + 2ab - \cancel{b^2}$$

$$= 2ab + 2ab = 4ab \qquad \text{[Proved]}$$

Now it is time to solve some easy mathematical problems.

**Example:**

Simplify:

$$999^2 + 2×999 + 1$$

Solution:

It will appear difficult when you try to simplify the expression using ordinary arithmetic. But when you apply one of the formulas, the simplification will be easy.

The expression can be expressed as,

$$999^2 + 2×999 + 1$$

$$= (999)^2 + 2×999×1 + (1)^2$$ [Applying the formula, $(a + b)^2 = a^2 + 2ab + b^2$ where a = 999 and b = 1]

$$= (999 + 1)^2$$

$$= (1000)^2$$

$$= 1000×1000$$

$$= 1000000$$ [Answer]

**Example:**

If  (a + b) = 4, and ab = 3

Determine the value of $a^2 + b^2$.

Solution:

We know,

$$a^2 + b^2 = (a + b)^2 - 2ab$$

Thus, if we determine the value of $(a + b)^2 - 2ab$ putting appropriate values, we get the answer.

$$a^2 + b^2 = (a + b)^2 - 2ab$$
$$= (4)^2 - 2 \times 3$$
$$= 16 - 6$$
$$= 10$$

Therefore, the value of $a^2 + b^2$ is 10.

**Example:**

Given: $x - \dfrac{1}{x} = 3$

Determine the value of,   (a)   $x^2 + \dfrac{1}{x^2}$

(b)   $x + \dfrac{1}{x}$

Solution:

(a)   We can expand the expression as,

$$x^2 + \dfrac{1}{x^2}$$

$$= \left( x - \dfrac{1}{x} \right)^2 + 2.x.\dfrac{1}{x} \qquad \text{[Applying the formula,}$$
$$a^2 + b^2 = (a - b)^2 + 2ab]$$

$$= \left( x - \dfrac{1}{x} \right)^2 + 2$$

$$= (3)^2 + 2 \qquad \text{[Putting } x - \dfrac{1}{x} = 3 \text{ ]}$$

$$= 9 + 2 = 11$$

Therefore, $x^2 + \dfrac{1}{x^2} = 11$

(b)

Let's consider the expression,

$$\left(x + \frac{1}{x}\right)^2$$

$$= x^2 + 2.x.\frac{1}{x} + \left(\frac{1}{x}\right)^2 \qquad \text{[Applying the formula,}\\ (a + b)^2 = a^2 + 2ab + b^2\text{]}$$

$$= x^2 + \frac{1}{x^2} + 2$$

$$= \left[\left(x - \frac{1}{x}\right)^2 + 2.x.\frac{1}{x}\right] + 2 \qquad \text{[Applying the formula,}\\ a^2 + b^2 = (a - b)^2 + 2ab\text{]}$$

$$= \left[\left(x - \frac{1}{x}\right)^2 + 2\right] + 2$$

$$= [(3)^2 + 2] + 2 \qquad \text{[Putting } x - \frac{1}{x} = 3 \text{ ]}$$

$$= 9 + 2 + 2 = 13$$

Therefore we get the following equation,

$$\left(x + \frac{1}{x}\right)^2 = 13$$

If we take square root of both sides, we get,

$$\left(x + \frac{1}{x}\right) = \pm\sqrt{13}$$

Therefore the value of $\left(x + \frac{1}{x}\right)$ is $\pm\sqrt{13}$.

Notice that here we wrote the square root of 13 as $\pm\sqrt{13}$ .

---

When we talk about the square root of any number, we should consider negative sign along with positive sign. For example, square root of 4 is ±2. It is quite obvious because,

(+2)×(+2) = 4 $\qquad\qquad$ (-2)×(-2) = 4

---

**Example:**

Given:    $a + b = 3$
          $ab = 2$

Determine the value of,    $a^4 + b^4$

Solution:

The expression can be expanded as,

$$a^4 + b^4$$

$$= (a^2)^2 + (b^2)^2$$

$$= (a^2 + b^2)^2 - 2a^2b^2$$

$$= [(a + b)^2 - 2ab]^2 - 2a^2b^2$$

[Applying the formula,
$a^2 + b^2 = (a + b)^2 - 2ab$]

$$= [(a + b)^2 - 2ab]^2 - 2(ab)^2$$

$$= [(3)^2 - 2 \times 2]^2 - 2 \times 2^2$$

[Putting
$a + b = 3$, $ab = 2$]

$$= [9 - 4]^2 - 2 \times 4$$

$$= 5^2 - 8 = 25 - 8 = 17$$

Therefore,

$$a^4 + b^4 = 17$$

In the previous section, we discussed about the formulas which are associated with square terms. Now we are going to discuss the formulas associated with cubes.

**Formula 6**

$$(a + b)^3 = a^3 + 3a^2b + 3ab^2 + b^3$$

**Formula 7**

$$(a - b)^3 = a^3 - 3a^2b + 3ab^2 - b^3$$

**Formula 8**

$$a^3 + b^3 = (a + b)(a^2 - ab + b^2)$$

$$= (a + b)^3 - 3ab(a + b)$$

**Formula 9**

$$a^3 - b^3 = (a - b)(a^2 + ab + b^2)$$

$$= (a - b)^3 + 3ab(a - b)$$

## Proof of formula 6

$(a + b)^3$ is the product of three $(a + b)$.

Thus,

$(a + b)^3$

$= (a + b)(a + b)(a + b)$

$= (a + b)[(a + b)^2]$     [ We know,

$= (a + b)(a^2 + 2ab + b^2)$    $a^2 + b^2 + 2ab = (a + b)^2$]

$= a^2(a + b) + 2ab(a + b) + b^2(a + b)$

$= a^3 + a^2b + 2a^2b + 2ab^2 + ab^2 + b^3$

$= a^3 + 3a^2b + 3ab^2 + b^3$    [Proved]

## Proof of formula 7

The proof of formula 7 is quite similar to the previous one.

$(a - b)^3$

$= (a - b)(a - b)(a - b)$

$= (a - b)[(a - b)^2]$

$= (a - b)(a^2 - 2ab + b^2)$    [ We know,

     $a^2 + b^2 - 2ab = (a - b)^2$]

$= a^2(a - b) - 2ab(a - b) + b^2(a - b)$

$= a^3 - a^2b - 2a^2b + 2ab^2 + ab^2 - b^3$

$= a^3 - 3a^2b + 3ab^2 - b^3$    [Proved]

## Proof of formula 8

We have to prove the following equation,

$$a^3 + b^3 = (a + b)(a^2 - ab + b^2)$$
$$= (a + b)^3 - 3ab(a + b)$$

We begin with the following expression,

$$a^3 + b^3$$
$$= a^3 + b^3 + 3a^2b + 3ab^2 - 3a^2b - 3ab^2$$

Since the overall value of $(3a^2b + 3ab^2 - 3a^2b - 3ab^2)$ is zero, the expression does not alter its value.

$$a^3 + b^3$$
$$= (a^3 + b^3 + 3a^2b + 3ab^2) - 3a^2b - 3ab^2$$
$$= (a + b)^3 - 3ab(a + b) \qquad \text{[Using the formula 6]}$$

Therefore,

$$a^3 + b^3 = (a + b)^3 - 3ab(a + b)$$

One part of the equation is proved. Now we should prove the remaining part.

$$a^3 + b^3 = (a + b)^3 - 3ab(a + b)$$
$$= (a + b)[(a + b)^2 - 3ab]$$
$$= (a + b)[a^2 + b^2 + 2ab - 3ab]$$
$$= (a + b)(a^2 - ab + b^2)$$

That is how we proved the remaining part.

# Proof of formula 9

$$a^3 - b^3$$

Just like the previous proof, we introduce the terms - $3a^2b$ + $3ab^2$, and in order to keep the value unaltered, we should also introduce +$3a^2b$ - $3ab^2$.

Thus,

$$a^3 - b^3$$
$$= (a^3 - b^3 - 3a^2b + 3ab^2) + 3a^2b - 3ab^2$$
$$= (a - b)^3 + 3ab(a - b) \qquad \text{[Using the formula 7]}$$

Proved part 1

$$a^3 - b^3 = (a - b)^3 + 3ab(a - b)$$
$$= (a - b)[(a - b)^2 + 3ab]$$
$$= (a - b)[a^2 - 2ab + b^2 + 3ab]$$
$$= (a - b)(a^2 + ab + b^2)$$

Proved part 2

Now let's discuss some examples.

**Example:**

Given:  $a + b = 5$

$ab = 6$

Determine the value of, $a^3 + b^3$

Solution:

Let's expand $(a^3 + b^3)$ according to the formula as,

$a^3 + b^3$

$= (a + b)^3 - 3ab(a + b)$

$= (5)^3 - 3 \times 6 \times 5 = 125 - 90 = 35$ [Answer]

**Example:**

If $x - \dfrac{1}{x} = 3$

Then

$$x^3 - \dfrac{1}{x^3} = ?$$

Solution:

$x^3 - \dfrac{1}{x^3}$

$= (x)^3 - \left(\dfrac{1}{x}\right)^3$

$= \left(x - \dfrac{1}{x}\right)^3 + 3.x.\dfrac{1}{x}.\left(x - \dfrac{1}{x}\right)$    Applying the formula, $a^3 - b^3 = (a - b)^3 + 3ab(a - b)$

$= \left(x - \dfrac{1}{x}\right)^3 + 3.\left(x - \dfrac{1}{x}\right)$

$= (3)^3 + 3 \times 3$

$= 27 + 9$

$= 36$      [Answer]


**Example:**

Given:    $a + b = 3$

          $ab = 2$

Determine the value of,

$$a^6 + b^6$$

Solution:

We may expand the expression as,

$a^6 + b^6$

$= (a^3)^2 + (b^3)^2$

$= (a^3 + b^3)^2 - 2a^3b^3$    [Applying the formula, $a^2 + b^2 = (a + b)^2 - 2ab$]

$= (a^3 + b^3)^2 - 2a^3b^3$

$= [(a + b)^3 - 3ab(a + b)]^2 - 2(ab)^3$

> Applying the formula,
> $a^3 + b^3 = (a + b)^3 - 3ab(a + b)$

Now we may put the values of $(a + b)$ and $ab$ in the above expression.

$= [(a + b)^3 - 3ab(a + b)]^2 - 2(ab)^3$

$= [(3)^3 - 3×2×3]^2 - 2×(2)^3$

$= [27 - 18]^2 - 2×8$

$= [9]^2 - 16$

$= 81 - 16$

$= 65$       [Answer]

## Example:

Multiply:

$$(2x + 3y)(4x^2 - 6xy + 9y^2)$$

Solution:

We may write the expression in the following way,

$$(2x + 3y)(4x^2 - 6xy + 9y^2)$$

$$= (2x + 3y)[(2x)^2 - 2x.3y + (3y)^2]$$

Now you can see the expression takes the form of the following formula.

$$a^3 + b^3 = (a + b)(a^2 - ab + b^2)$$

$$\text{Where } a = 2x \text{ and } b = 3y$$

Applying the formula we may reduce the given expression as,

$(2x + 3y)[(2x)^2 - 2x.3y + (3y)^2]$

$= (2x)^3 + (3y)^3$

$= 8x^3 + 27y^3$      [Answer]

# EXERCISE 4

1) Solve the following multiplication problems:

   (a)   $a(x + y)$                     (h)   $(a + b)(x + y)$

   (b)   $3(p + q)$                     (i)   $(2a + 3b)(2x - 3y)$

   (c)   $2(a + 3)$                     (j)   $(x + 3)(x + 1)$

   (d)   $4(2a - b)$                    (k)   $(2x + 4)(3x - 7)$

   (e)   $x(x + a)$                     (l)   $(x - 3a)(2x + 5a)$

   (f)   $3x(2x - 5)$                   (m)   $(2x - a)(a - 3x)$

   (g)   $3a(2p + 3q)$                  (n)   $(a + 1)(a + 2)(a + 3)$

2) Solve the following multiplication problems:

   (a)   $(x + a)(x - a)$

   (b)   $(2x + a)(2x - a)$

   (c)   $(2p + 3q)(2p - 3q)$

   (d)   $(a + b)^2(a - b)$

   (e)   $(a + b)^2(a - b)^2$

   (f)   $(x - a)(x^2 + xa + a^2)$

   (g)   $(2x + a)(4x^2 - 2xa + a^2)$

   (h)   $(5x - 3y)(25x^2 + 15xy + 9y^2)$

3) Solve the following multiplication problems:

   (a)   $(a + b)(a + b + c)$

   (b)   $(a + b)(ap^2 + bq^2)$

   (c)   $(a + b)(c + d)(e + f)$

   (d)   $(2x + 3)(x^2 - 3x + 6)$

   (f)   $(a^2 + 2ab + b^2)(a^2 - 2ab + b^2)$

4)      Simplify the following expressions.

(a)     $x^2 - x(x + a)$

(b)     $x(x + a) - x(x - a)$

(c)     $(x + 3)(x + 2) - (x + 6)(x - 1)$

(d)     $x(x + 4) + x(x + 3) - x(x - 2)$

(e)     $(a + b)(a + 2) - a^2$

(f)     $\dfrac{2x - 6}{x - 3}$ [Here x - 3 ≠ 0]

(g)     $a(b + c) - b(a - c) + c(a - b)$

(h)     $\dfrac{2a + 2}{a + 1} + \dfrac{3a + 6}{a + 2}$    [Here a + 1 ≠ 0 and a + 2 ≠ 0]

(i)     $a - [a - \{a - (a - 1)\}]$

(j)     $\dfrac{2x}{x + 1} + \dfrac{2}{x + 1}$    [Here x + 1 ≠ 0]

(k)     $\dfrac{1}{a - b} + \dfrac{1}{b - a}$    [Here a ≠ b]

5) Prove the following relations:

(i)     $(a + b + c)^2 = a^2 + b^2 + c^2 + 2ab + 2bc + 2ca$

(ii)    $a^3 + b^3 + c^3 - 3abc = (a + b + c)(a^2 + b^2 + c^2 - ab - bc - ca)$

(iii)   $(a + b - c)^2 = a^2 + b^2 + c^2 + 2ab - 2bc - 2ca$

(iv)    $(a - b - c)^2 = a^2 + b^2 + c^2 - 2ab + 2bc - 2ca$

(v)     $(a - b)^2 = (a + b)^2 - 4ab$

(vi)    $(a + b)^2 = (a - b)^2 + 4ab$

(vii)   $a^4 - b^4 = (a + b)(a - b)(a^2 + b^2)$

6)  If $a + b = 2$, determine the values of the following expressions.

      (i)    $a^2 + 2ab + b^2$

      (ii)   $a^3 + 3a^2b + 3ab^2 + b^3$

7)  If $x - 2y = 3$, determine the values of the following expression.

$$x^2 - 4xy + 4y^2$$

8)  If $x + \dfrac{1}{x} = 5$, determine the values of the following expressions.

      (i)  $x^2 + \dfrac{1}{x^2}$

      (ii)  $x^3 + \dfrac{1}{x^3}$

      (iii)  $x^4 + \dfrac{1}{x^4}$

9)  Simplify the following expression applying formula.

$$(2.01)^2 - 2 \times 2.01 \times 0.01 + (0.01)^2$$

10)  Given, $a - b = 1$

$$ab = 12$$

Determine the values of the following expressions.

      (i)    $a^2 + b^2$

      (ii)   $a^3 - b^3$

      (iii)  $a + b$

11) If $a + b + c = 0$, prove that,
$$a^3 + b^3 + c^3 = 3abc$$

12) Given $x + \dfrac{1}{x} = 2$, then determine the value of,

$$x^{31} + \dfrac{1}{x^{31}}$$

13) Given $a + b + c = 1$

$$a^2 + b^2 + c^2 = 2$$

$$a^3 + b^3 + c^3 = 3$$

Then prove that,

$$a^4 + b^4 + c^4 = 4\dfrac{1}{6}$$

14) Given $x + \dfrac{1}{x} = 3$,

Determine the value of,

$$x^5 + \dfrac{1}{x^5}$$

# CHAPTER 5

# INDICES OR POWERS

Along with simple addition, subtraction, multiplication, and division, you should also learn how to deal with indices (or powers) in order to solve several algebra problems.

We already discussed when a term is multiplied with itself several times, we represent it with power form.

For example,

$a.a.a$ is represented as $a^3$.

$$a.a.a = a^3$$

The meaning of the following expressions is shown below:

$x^4$ means the product of four $x$.

$$x^4 = x.x.x.x$$

$a^7$ means the product of seven $a$.

$$a^7 = a.a.a.a.a.a.a$$

When the product of two or more terms is raised to its power, how do you represent it?

Suppose the product of a and b is raised to its fifth power.

$$(ab)^5$$

The above expression means the product of a and b is multiplied five times.

$$(ab).(ab).(ab).(ab).(ab)$$

We may alternatively write the expression as,

$$(ab).(ab).(ab).(ab).(ab)$$

$$= a.b.a.b.a.b.a.b.a.b$$

$$= a.a.a.a.a.b.b.b.b.b$$

$$= (a.a.a.a.a).(b.b.b.b.b)$$

$$= (a^5)(b^5)$$

$$= a^5b^5$$

Therefore, $(ab)^5$ can be alternatively represented as, $a^5b^5$.

$$(ab)^5 = a^5b^5$$

Consider another expression.

$$(2a)^3$$

You know that $2a$ represents the product of 2 and $a$. Thus you may represent the expression as,

$$(2a)^3$$

$$= 2^3a^3$$

$$= 8a^3 \qquad \text{[Since } 2^3 = 8.\text{]}$$

Therefore when powers are equal of different terms, and they are multiplied together, such as,

$$a^6b^6c^6$$

You may raise the power to their product as,

$$(abc)^6$$

**We may express it as a general formula as,**

$$a^m \cdot b^m = (ab)^m \longrightarrow (1)$$

However when powers are not equal, the above expression is not applicable. You should simply write them as,

$$a^m b^n \qquad x^2 y^3 \qquad pq^2 r^4$$

A similar equation will also be true in case of division.

$$\frac{a^m}{b^m} = \left(\frac{a}{b}\right)^m \longrightarrow (2)$$

Some examples:

$$\frac{2^4}{3^4} = \left(\frac{2}{3}\right)^4$$

$$\frac{5^7}{6^7} = \left(\frac{5}{6}\right)^7$$

$$\frac{1}{8} = \frac{1^3}{2^3} = \left(\frac{1}{2}\right)^3$$

Now consider the following expression,

$$a^5.a^3$$

Here for the first term, the power of a is 3, and for the second term, the power of a is 3. Thus it is the multiplication of two entities which are different powers of the same term.

Let's simplify the expression using our basic knowledge of mathematics.

$a^5.a^3$

$= (a.a.a.a.a).(a.a.a)$    [We know,
$a^5 = a.a.a.a.a$

$= a.a.a.a.a.a.a.a$    and

$a^3 = a.a.a$]

$= a^8$

Therefore,

$$a^5.a^3 = a^8$$

You should notice here that the number 8 is the sum of 5 and 3. Therefore, when such multiplication is given, you need to add their powers to find the answer.

A general rule can be represented as,

$$a^m.a^n = a^{m+n} \longrightarrow (3)$$

We may apply a similar technique in case of division too. Suppose the following expression is given,

$$\frac{a^5}{a^3} \quad (Or,\ a^5 \div a^3)$$

Let's simplify the expression.

$$\frac{a^5}{a^3}$$

$$= \frac{a.a.a.a.a}{a.a.a}$$

$$= a.a = a^2$$

Therefore,

$$\frac{a^5}{a^3} = a^2$$

If you carefully notice, you realize the number 2 comes from the subtraction of 3 from 5.

$$\frac{a^5}{a^3} = a^{5-3} = a^2$$

Therefore we may construct a general formula from the above discussion.

$$\frac{a^m}{a^n} = a^{m-n} \longrightarrow (4)$$

Applying the above formula, we are able to prove an interesting mathematical relation, which is,

$$a^0 = 1 \longrightarrow (5)$$

a to the power 0 is equal to 1, where may be any number except 0.

**Proof:**

We may write 0 as,

$$m - m = 0 \qquad \text{[Where m is any positive number.]}$$

Thus $a^0$ can be written as,

$$a^{m - m}$$

Since we know,

$$\frac{a^m}{a^n} = a^{m - n}$$

We may write $a^{m-m}$ as,

$$a^{m-m} = \frac{a^m}{a^m}$$

$$= 1 \quad \text{[Proved]}$$

Note that, the equation $a^0 = 1$ is not valid when $a = 0$, because if we considered $a = 0$, we would get,

$$0^{m-m} = \frac{0^m}{0^m} = \frac{0}{0} \quad \text{Which is undefined.}$$

## The meaning of Negative Power

We are going to prove another important relation, which is associated with negative power.

The relation is,

$$a^{-n} = \frac{1}{a^n} \longrightarrow (6)$$

In order to prove the above relation, we begin with the relation,

$$\frac{a^m}{a^n} = a^{m-n}$$

Let m = 0 in the above equation. Thus we replace m by 0.

$$a^{0-n} = \frac{a^0}{a^n}$$

$$\text{or,} \quad a^{-n} = \frac{1}{a^n} \quad \text{[Proved]}$$

For example, consider the following expression,

$$2^{-3}$$

We can also write this expression as,

$$2^{-3} = \frac{1}{2^3}$$

$$= \frac{1}{8}$$

134  Therefore, $2^{-3}$ and $1/8$ have the same meaning.

## Power of a power term

Consider the following term,

$$(a^3)^2$$

*It is square of $a^3$.*

You know that $a^3$ is obtained when a is raised to power 3.

And when the term $a^3$ is raised to power 2, we get $(a^3)^2$.

$$(a^3)^2$$

We may expand the above expression in the following way,

$$(a^3)^2$$
$$= (a^3).(a^3)$$
$$= (a.a.a).(a.a.a)$$
$$= a.a.a.a.a.a$$
$$= a^6$$

Therefore,

$$(a^3)^2 = a^6$$

You must understand here the power 6 arises from the product of 3 and 2. We may construct a general formula as,

$$(a^m)^n = a^{mn} \qquad \longrightarrow (7)$$

A few examples:

$$(x^3)^5 = x^{15}$$

$$(a^4)^3 = a^{12}$$

$$(5^2)^7 = 5^{14}$$

$$(3^5)^4 = 3^{20}$$

## Fractional Power

The following expression is given.

$$8^{\frac{1}{3}}$$

*8 to the power 1/3.*

Can you simplify the expression to determine its value?

[In order to simplify the expression, we need to solve an equation. However, you will learn the technique of solving equations in the next chapter. So, if you are unable to understand how the mathematical procedure is done, do not worry much, because you will understand very soon when you go through the next chapter.]

Consider the value of $8^{1/3}$ is equal to an unknown term x.

$$x = 8^{\frac{1}{3}}$$

Since it is an equation, we can perform identical operations on the both sides of the equation.

$$x = 8^{\frac{1}{3}}$$

$$\text{or, } x^3 = (8^{\frac{1}{3}})^3 \quad \text{[We raised the power of both sides to cube.]}$$

$$\text{or, } x^3 = 8^{\frac{1}{3} \cdot 3} \quad \text{[Since we know, } (a^m)^n = a^{mn}.]$$

$$\text{or, } x^3 = 8$$

$$\text{or, } x = \sqrt[3]{8} \quad \text{[Considering cube root for the both sides.]}$$

$$\text{or, } x = 2$$

Therefore the simplified value of the expression is 2.

$$8^{\frac{1}{3}} = 2$$

**Therefore, when a term is raised to the power $1/_m$, it actually represents m-th root of that term.**

For example,

$4^{\frac{1}{2}}$ means $\sqrt{4}$, which is equal to 2.

$16^{\frac{1}{4}}$ means $\sqrt[4]{16}$, which is equal to 2.

$27^{\frac{1}{3}}$ means $\sqrt[3]{27}$, which is equal to 3.

We may represent a general formula:

$$a^{\frac{1}{m}} = \sqrt[m]{a} \longrightarrow (8)$$

We can make a proposition with the help of above formula. Note that the fraction $1/_m$, which acts as the power, has 1 as the numerator. But what will be the case when numerator is other than 1?

Let's consider the following expression,

$$8^{\frac{2}{3}}$$

Here the power is $2/_3$, and it can be written as the product of $1/_3$ and 2.

Therefore,

$8^{\frac{2}{3}}$

$= 8^{\frac{1}{3} \times 2}$

$= (8^{\frac{1}{3}})^2 \quad$ [As $(a^m)^n = a^{mn}$]

$= (\sqrt[3]{8})^2$

$= (2)^2 = 4$

Therefore the general expression can be represented as,

$$a^{\frac{n}{m}} = (\sqrt[m]{a})^n \quad\longrightarrow (9)$$

Now it is time to show some examples:

**Example:**
Write the simplified forms of the following expressions.

(a)   $x^{12}.x^{12}$

(b)   $(x^5)^6$

(c)   $x^3.x^4.x^5$

(d)   $x^{23}.x^{-11}$

(e)   $\dfrac{x^{34}}{x^{21}}$

(f)   $\dfrac{x^5.x^8}{x^7}$

(g)   $x^{\frac{2}{5}}.x^{\frac{3}{5}}$

(h)   $\sqrt[3]{x^{15}}$

Solution:

(a)   $x^{12}.x^{12}$

   $= x^{(12 + 12)}$   [Applying the formula, $a^m.a^n = a^{m+n}$]

   $= x^{24}$        [Answer]

(b)   $(x^5)^6$

   $= x^{5 \times 6}$    [Applying the formula, $(a^m)^n = a^{mn}$]

   $= x^{30}$       [Answer]

(c)     $x^3 \cdot x^4 \cdot x^5$

$= x^{(3+4)} \cdot x^5$     [Applying the formula, $a^m \cdot a^n = a^{m+n}$]

$= x^7 \cdot x^5$

$= x^{(7+5)}$

$= x^{12}$     [Answer]

(d)     $x^{23} \cdot x^{-11}$

$= x^{23} \cdot \dfrac{1}{x^{11}}$     [Applying the relation, $a^{-n} = \dfrac{1}{a^n}$]

$= \dfrac{x^{23}}{x^{11}}$

$= x^{(23-11)}$     [Applying the formula, $\dfrac{a^m}{a^n} = a^{m-n}$]

$= x^{12}$     [Answer]

(e)     $\dfrac{x^{34}}{x^{21}}$

$= x^{(34-21)}$     [Applying the formula, $\dfrac{a^m}{a^n} = a^{m-n}$]

$= x^{13}$     [Answer]

(f)     $\dfrac{x^5 \cdot x^8}{x^7}$

$= \dfrac{x^{(5+8)}}{x^7}$     [Applying the formula, $a^m \cdot a^n = a^{m+n}$]

                     [Applying the formula, $\dfrac{a^m}{a^n} = a^{m-n}$]

$= \dfrac{x^{13}}{x^7} = x^{(13-7)} = x^6$ [Answer]

**(g)**

$$x^{\frac{2}{5}} \cdot x^{\frac{3}{5}}$$

$$= x^{\left(\frac{2}{5} + \frac{3}{5}\right)} \qquad \text{[Applying the formula, } a^m.a^n = a^{m+n}]$$

$$= x^{\frac{2+5}{5}}$$

$$= x^{\frac{5}{5}} = x^1 = x \qquad \text{[Answer]}$$

**(h)**

$$\sqrt[3]{x^{15}}$$

$$= \left(x^{15}\right)^{\frac{1}{3}} \qquad \text{[Applying the relation } a^{\frac{1}{m}} = \sqrt[m]{a} \text{ ]}$$

$$= x^{\frac{15}{3}} \qquad \text{[Applying the formula, } (a^m)^n = a^{mn}]$$

$$= x^5 \qquad \text{[Answer]}$$

## Example:

Multiply:

$$\left(x^3 + \frac{1}{x^3}\right)\left(x^2 + \frac{1}{x^2}\right)$$

Solution:

$$\left(x^3 + \frac{1}{x^3}\right)\left(x^2 + \frac{1}{x^2}\right)$$

$$= x^2\left(x^3 + \frac{1}{x^3}\right) + \frac{1}{x^2}\left(x^3 + \frac{1}{x^3}\right)$$

$$= x^2.x^3 + x^2.\frac{1}{x^3} + \frac{1}{x^2}.x^3 + \frac{1}{x^2}.\frac{1}{x^3}$$

$$= x^2.x^3 + \frac{x^2}{x^3} + \frac{x^3}{x^2} + \frac{1}{x^2.x^3}$$

$$= x^5 + \frac{1}{x} + x + \frac{1}{x^5} \qquad \text{[Answer]}$$

# EXERCISE 5

1) Simplify the following expressions.

(a) $(2^2)^2$

(b) $\dfrac{7^6}{7^4}$

(c) $\left(\dfrac{27}{8}\right)^{-\frac{2}{3}}$

(d) $\left[\left(\dfrac{3^7}{3^3}\right)\cdot\left(\dfrac{2^{11}}{2^7}\right)\right]^{\frac{1}{2}}$

2) Simplify the following expressions.

(a) $(x^5y^7)\cdot(x^3y^{-2})$

(b) $\dfrac{x^7y^3}{x^5y^5}$

(c) $\left[(x^7y^3)(x^3y^7)\right]^{\frac{1}{5}}$

(d) $(a^{-1}+b^{-1})^{-1}(a+b)$

3) If $x^5 = -1$, tell one possible value of $x$.

4) Simplify:
$$(2^x + 2^{-x})(3^x + 3^{-x})$$

5) Given $f(x) = 5^x + \dfrac{1}{5^x} + 2$

Prove that $f(-x) = f(x)$

5) Prove the following relation.

$$\left(\dfrac{x^a}{x^b}\right)^c \left(\dfrac{x^b}{x^c}\right)^a \left(\dfrac{x^c}{x^a}\right)^b = 1$$

# CHAPTER 6

# EQUATIONS AND THEIR SOLUTIONS

You already know the meaning of a equal sign (=). This sign is generally used to show that two expressions have equal values, however they look different.

As an example, consider these two expressions.

$$(7 + 4) \text{ and } (19 - 8)$$

These two expressions consist of different numbers and different mathematical signs. But when you calculate their values, you find they are equal, which is 11.

Therefore, you may write an equal sign (=) between them expressing as the following equation.

$$7 + 4 = 19 - 8$$

However, the above equation does not tell any new thing to us. We may write an infinite number of such equations, but they do not have any application to solve a mathematical problem.

Interestingly, when we apply the knowledge of algebra to an equation, several difficult mathematical problems can be solved easily.

Let's start with a simple problem, and discuss how we can alternatively use algebra to solve it.

Your friend Bob has some fishes in his aquarium, and you have 7 fishes in your aquarium. Total number of fishes of you and Bob is 15. How many fishes does Bob have?

The solution of the problem is quite simple. It will be easy to solve the problem using your basic knowledge of arithmetic.

You know that the total number of fishes of you and Bob is 15.

And you also know that you have 7 fishes.

Therefore, you can easily calculate the number of fishes in Bob's aquarium subtracting 7 from 15.

$$15 - 7 = 8$$

The number of fishes in Bob's aquarium is 8.

**Now we are going to learn an alternative way to solve the problem using algebra.**

At this point, you know a little algebra, and also you know that an unknown quantity can be represented by a letter in place of a number. In the given math problem, the unknown quantity is the number of Bob's fishes. Thus we may represent the number of Bob's fishes with a letter.

Let the number of Bob's fishes is x.

Since you know you have 7 fishes in your aquarium, we may represent the total number of fishes by the following expression.

$$x + 7$$

On the other hand, the total number of fishes is already given. The total number of fishes of you and Bob is 15.

Therefore, we may construct the following equation.

$$x + 7 = 15$$

Now we should solve the equation in order to determine the value of x.

$$x + 7 = 15$$

The left hand side of the equation is the sum of x and 7. **Our aim is to remove all other terms from the left hand side except x, in a way that the final form becomes 'x is equal to a number'.**

Let's discuss how we can do it.

$$x + 7 = 15$$

Since it is an equation, it will remain valid if we add or subtract the same quantity on both sides of the equation.

In the present case, we need to subtract 7 from the left hand side of the equation in order to free the x. In order to keep the equation valid, we have to subtract 7 from the right hand side also.

Therefore, the equation becomes,

$$x + 7 - 7 = 15 - 7$$   (Subtracting 7 from both sides of the equation)

or, $x = 8$

Thus we get x is equal to 8 as the solution of the equation.

Since x represents the number of Bob's fishes, we can say Bob has 8 fishes in his aquarium.

**But why would someone use algebra to solve a math problem, where arithmetic is also able to solve the math problem efficiently?**

You will understand the reason when you try to solve the following math problem using arithmetic.

Suppose you received 5 packs of candies from your teacher, and he instructed you to distribute the candies among all the students. All the packs contained equal numbers of candies. You poured all the candies into a bowl opening the packs, and then gave 3 candies to each student. There were 20 students in the classroom. After the distribution, you found that 10 still left in the bowl. Now determine the number of candies initially in every pack.

This math problem is quite difficult to solve using ordinary arithmetic. So we should use algebra in order to solve such problems.

Let the number of candies in each pack was x.

There were 5 packs of candies. So the total number of candies was 5x.

$$5x$$

Since 10 candies remained in the bowl after distribution, the number of distributed candies was,

$$5x - 10$$

Since there were 20 students in the classroom, each student received,

$$\frac{5x - 10}{20} \text{ candies.}$$

On the other hand, it is already given that each student received 3 candies.

Therefore we construct the following equation,

$$\frac{5x - 10}{20} = 3$$

When we solve the above equation, we will get the value of x, which represents the number of candies in each pack. But before solving the equation, we need to know some rules, which are applicable for any equation.

- **We can add equal terms to both sides of an equation.**

- **We can subtract equal terms from both sides of an equation.**

- **We can multiply both sides of an equation by equal terms except zero.**

- **We can divide both sides of an equation by equal terms except zero.**

Besides addition, subtraction, multiplication, and division, there are a few more operations which can be performed to the both sides of an equation. For example, both sides can be raised to a definite power. When you will study some advanced algebra books, you will get to know about them.

Now let's try to solve the equation which we constructed. You need to follow carefully how we are going to free x from all other terms at the left hand side of the equation.

$$\frac{5x - 10}{20} = 3$$

Notice, the left hand side is associated with denominator 20. Thus if we multiply 20 on both sides, we get a simplified form.

$$\frac{5x - 10}{20} = 3$$

or, $\frac{5x - 10}{20} \times 20 = 3 \times 20$ (Multiplying both sides by 20)

or, $5x - 10 = 60$

In the next step, we should add 10 to both sides of the equation.

$$5x - 10 = 60$$

or, $5x - 10 + 10 = 60 + 10$ (Adding 10 to both sides)

or, $5x = 70$

Now we should divide both sides by 5 in order to isolate x.

$$5x = 70$$

or, $\dfrac{5x}{5} = \dfrac{70}{5}$ (Dividing both sides by 5)

or, $x = 14$

We get 14 as the value of x. Therefore, we conclude each pack contained 14 candies.

Let's deal with a few more examples.

**Example:**

Solve for x.

$$3x + 7 = 2x + 5$$

Solution:

Both sides of the equation have the terms containing x. We need to bring all x containing terms to the left side.

$$3x + 7 = 2x + 5$$

or, $3x + 7 - 2x = 2x + 5 - 2x$ (Subtracting 2x from both sides)

or, $x + 7 = 5$

In order to isolate x at the left hand side, we should subtract 7 from both sides of the equation.

$$x + 7 = 5$$

$$\text{or, } x + 7 - 7 = 5 - 7$$

(Subtracting 7 from both sides)

$$\text{or, } x = -2$$

Therefore the value of x is - 2.

**Example:**

Solve for x.

$$\frac{3x + 1}{x + 2} = 2$$

Solution:

$$\frac{3x + 1}{x + 2} = 2$$

$$\text{or, } 3x + 1 = 2(x + 2)$$    [Multiplying both sides by (x + 2)]

$$\text{or, } 3x + 1 = 2x + 4$$

$$\text{or, } x + 1 = 4$$    (Subtracting 2x from both sides)

$$\text{or, } x = 3$$    (Subtracting 1 from both sides)

Therefore the value of x is 3.

**Example:**

The sum of three consecutive integers is 51. Determine those integers.

Solution:

Let the smallest integer is x.

Thus the next integer should be (x + 1).

And the largest integer is (x + 2).

Thus the sum of these three consecutive integers can be expressed as,

$$x + (x + 1) + (x + 2)$$

$$= 3x + 3$$

Given in this problem, the sum of three consecutive is 51. Therefore, we may construct an equation, and solve it.

$$3x + 3 = 51$$

or, $3x = 51 - 3$      (Subtracting 3 from both sides)

or, $3x = 48$

or, $x = 48/3$      (Dividing both sides by 3)

or, $x = 16$

Since x = 16, the smallest integer is 16

Thus,

$$x + 1 = 17$$

$$x + 2 = 18$$

Therefore, the three consecutive integers are, 16, 17, and 18.

**Example:**

Express the following repeating decimal as the ratio of two whole numbers.

0.83333333333...

Solution:

You already know from chapter 1 how to express a repeating decimal into its fractional form of the ratio of two whole numbers.

However you are able to apply the previous method when same set of numbers are repeating after the decimal point. But in the present case, you notice the digit just after the decimal point is 8, and then 3s are repeating. So you need to apply algebra in such cases. Let's see how.

Let,

$$x = 0.833333333...$$

or, $\quad 10x = 8.33333333....$ [Multiplying both sides by 10.]

or, $\quad 10x = 8 + 0.3333333...$

or, $\quad 10x = 8 + \dfrac{1}{3}$

or, $\quad 10x = \dfrac{25}{3}$

or, $\quad x = \dfrac{25}{30}$

or, $\quad x = \dfrac{5}{6}$

Since we considered x = 0.8333333..., and we get x = 5/6, we may conclude that 0.833333... = 5/6 (which is a ratio of two whole numbers).

# EXERCISE 6

1) Solve for x.

   (a)   $x + 3 = 0$

   (b)   $x - 5 = 2$

   (c)   $2x = 10$

   (d)   $\dfrac{x}{3} = 5$

   (e)   $\dfrac{2}{x} = 10$

   (f)   $2x + 3 = 11$

   (g)   $\dfrac{4x + 7}{3} = 5$

   (h)   $\dfrac{2x + 3}{3x + 5} = 1$

   (i)   $\dfrac{x}{2} + \dfrac{x}{3} = 5$

   (j)   $\dfrac{x + 1}{3} + \dfrac{x - 1}{2} = 1$

   (k)   $(x + 1)(x + 2) = (x + 3)(x + 4)$

2) Express the following repeating decimals as ratio of whole numbers.

   (a)   0.1666666666...

   (b)   0.91666666666...

   (c)   1.1666666666...

3) The difference of two numbers is 9. If the smaller number is 11, determine the larger number.

4) The sum of two numbers is 23. One number is 11, determine the other number.

5) When 3 is added to the twice of a number, we get 45. Determine the number.

6) The sum of five times and two times of a number is 49. What is the number?

7) The sum of present ages of father and son is 70 years. The son was born when the father was 24 years old. Determine the present age of the son.

8) You have double number of books than Bob. Total number of books of you and Bob is 30. How many books does Bob have?

9) When you add 7 to the five times of a number, you get 32. What number is it?

10) The sum of two consecutive numbers is 67. Determine those numbers.

11) Distribute 300 dollars between Sam and Harry in a way that Sam gets 100 dollars more than Harry.

12) The sum of two numbers is 60. One number is double of another. Determine these numbers.

13) Prove that the sum of three consecutive integers is always divisible by 3.

14) The sum of three consecutive even numbers is 72. What are those numbers?

15) The sum and the difference of two numbers are 44 and 12 respectively. Determine the numbers.

16) The sum of $^1/_3$ part and $^2/_5$ part of a number is 44. Determine the number.

17) The sum of present ages of father and son is 70 years. Five years ago, the father's age was five times of his son. Determine the present ages of father and his son.

# CHAPTER 7

# FACTORIZATION

Most of the integers are perfectly divisible by some other smaller integers. Those smaller integers are known as factors of that large integer.

For example, 6 is perfectly divisible by 2 and 3. Thus 2 and 3 are factors of 6.
30 is divisible by 2, 3, and 5. Thus 2, 3, and 5 are factors of 30.

**The method of representing an integer as the product form of its respective factors is known as factorization.**

For example, you can factorize the following integers as,

$$6 = 2 \times 3$$

$$9 = 3 \times 3 \times 3$$

$$12 = 2 \times 2 \times 3$$

$$15 = 3 \times 5$$

$$20 = 2 \times 2 \times 5$$

The above discussion is not new to you. It is quite expected that you already learned in your school how to factorize a number.

In this chapter, we are going to discuss how an algebraic expression can be represented as the product of its respective factors.

Let's consider the following expression,

$$3axy$$

This expression is the product of 3, a, x, and y.

As you can see the expression is already in its factorized form, there is no need to perform any mathematical treatment to factorize it further.

Now consider the following expression,

$$3axy + y$$

This expression is the sum of two terms, 3axy and y. Since y is common in both the terms, we may reduce the expression as.

$$3axy + y$$

$$= y(3ax + 1)$$

The expression became the product of two terms, y and (3ax + 1), and it can not be factorized further. Therefore the factorization of (3axy + y) yields y(3ax + 1).

Let's try another expression.

$$xy + ay + bx + ab$$

The given expression is the sum of four terms, xy, ay, bx, and ab. Now carefully follow how the given expression is being factorized.

$$xy + ay + bx + ab$$

$$= y(x + a) + b(x + a)$$

$$= (x + a)(y + b)$$

Thus the factorized form is,

$$(x + a)(y + b)$$

**Hopefully you have noticed that factorization of any algebraic expression is merely the reverse treatment of multiplication.**

In some cases, we may apply the formulas which you learned in chapter 4.

Consider the following formula.

$$a^2 - b^2 = (a + b)(a - b)$$

Here the expression is just expressed as the product of factors, $(a + b)(a - b)$. Thus when we find an expression similar to $(a^2 - b^2)$, we may apply the formula to factorize it.

**Example:**

Factorize:

$$4x^2 - 9y^2$$

Solution:

The given expression can be expressed as,

$$4x^2 - 9y^2$$

$$= (2x)^2 - (3y)^2$$

This expression is equivalent to $(a^2 - b^2)$, where we may consider $2x$ as $a$, and $3y$ as $b$.
Thus the factorized form is,

$$(2x)^2 - (3y)^2$$

$$= (2x + 3y)(2x - 3x) \quad \text{[Answer]}$$

**Example:**

Factorize:

$$a^2 - 1$$

Solution:

$$a^2 - 1$$

$$= (a)^2 - (1)^2$$

$$= (a + 1)(a - 1)$$

Therefore the factorized form is,

$$(a + 1)(a - 1)$$

**Example:**

Factorize:

$$x^4 - y^4$$

Solution:

$$x^4 - y^4$$

$$= (x^2)^2 - (y^2)^2$$

$$= (x^2 + y^2)(x^2 - y^2) \quad \text{[Further factorization is possible.]}$$

$$= (x^2 + y^2)(x + y)(x - y) \quad \text{[Answer]}$$

**Example:**

Factorize:

$$a^4 + a^2 + 1$$

Solution:

The method of factorization of the given expression a bit tricky.

We may expand the expression as,

$$a^4 + a^2 + 1$$

| | |
|---|---|
| $= a^4 + 2a^2 - a^2 + 1$ | [ We expressed the term $a^2$ as, $(2a^2 - a^2)$.] |
| $= a^4 + 2a^2 + 1 - a^2$ | |
| $= [(a^2)^2 + 2.\ a^2.1 + (1)^2] - a^2$ | [We know the formula, $(a + b)^2 = a^2 + 2ab + b^2$] |
| $= (a^2 + 1)^2 - a^2$ | |
| $= (a^2 + 1 + a)(a^2 + 1 - a)$ | [Applying the formula, $a^2 - b^2 = (a + b)(a - b)$] |

[Answer]

The following two formulas are also applicable in order to factorize some expressions.

- $a^3 + b^3 = (a + b)(a^2 - ab + b^2)$

- $a^3 - b^3 = (a - b)(a^2 + ab + b^2)$

Let's factorize a few expressions applying above formulas.

**Example:**

Factorize:

$$x^3 + 8$$

Solution:

$x^3 + 8$

$= x^3 + (2)^3$

$= (x + 2)[x^2 - 2.x + (2)^2]$ [Applying the formula:
$a^3 + b^3 = (a + b)(a^2 - ab + b^2)$]

$= (x + 2)(x^2 - 2x + 4)$

[Answer]

**Example:**

Factorize:

$27 - x^3$

Solution:

$27 - x^3$

$= (3)^3 - x^3$

$= (3 - x)[(3)^2 + (3).(x) + x^2]$ [Applying the formula:
$a^3 - b^3 = (a - b)(a^2 + ab + b^2)$]

$= (3 - x)(9 + 3x + x^2)$

[Answer]

The following formulas are also associated factorization of some expressions.

$$a^2 + 2ab + b^2 = (a + b)^2$$

$$a^2 - 2ab + b^2 = (a - b)^2$$

$$a^3 + 3a^2b + 3ab^2 + b^3 = (a + b)^3$$

$$a^3 - 3a^2b + 3ab^2 - b^3 = (a - b)^3$$

The right hand sides of these four equations represent the factorized form of the left hand sides. The squares or cubes of the expressions (a + b) or (a - b) are nothing but (a + b) or (a - b) multiplied two times or three times.

$$(a + b)^2 = (a + b)(a + b)$$

$$(a + b)^3 = (a + b)(a + b)(a + b)$$

**Example:**

Factorize:

$$x^2 + \frac{1}{x^2} + 2$$

Solution:

$$x^2 + \frac{1}{x^2} + 2$$

$$= (x)^2 + \left(\frac{1}{x}\right)^2 + 2.x.\frac{1}{x} \qquad \text{[We know the formula,}$$
$$(a + b)^2 = a^2 + 2ab + b^2]$$

$$= \left(x + \frac{1}{x}\right)^2$$

$$= \left(x + \frac{1}{x}\right)\left(x + \frac{1}{x}\right) \qquad \text{[Answer]}$$

**Example:**

Factorize:

$$a^3 - 6a^2 + 12a - 8$$

Solution:

$$a^3 - 6a^2 + 12a - 8$$

$$= (a)^3 - 3.a^2.2 + 3.a.(2)^2 - (2)^3 \qquad \text{[We know the formula:}$$
$$a^3 - 3a^2b + 3ab^2 - b^3$$
$$= (a - b)^3]$$

$$= (a - 2)^3$$

$$= (a - 2)(a - 2)(a - 2) \text{ [Answer]}$$

# Factorization through Splitting of the Middle Term

In order to factorize some expressions, we need to apply some special techniques. Splitting of the middle term is one of such techniques. Let's discuss the method depicting a few examples.

### Factorization of the following form

$$x^2 + bx + c$$

Where b and c are integers.

Let's consider the following multiplication problem.

$$(x + 2)(x + 3)$$

If you carry out the multiplication, you get,

$$(x + 2)(x + 3)$$

$$= x(x + 2) + 3(x + 2)$$

$$= x^2 + 2x + 3x + 6$$

$$= x^2 + 5x + 6$$

The resulting expression is similar to the expression, $x^2 + bx + c$, where b and c are 5 and 6 respectively.

**We already know that factorization is just the reverse operation of multiplication.**

**It means that the following expression will be given,**

$$x^2 + 5x + 6$$

**And you have to express it as,**

$$(x + 2)(x + 3)$$

Let's explain how we can do it.

$$x^2 + 5x + 6$$

## Step 1

Consider the third term which is not associated with x. Then express the term as the multiplied form of its possible factors.

For the given expression, the third term is 6. Now 6 can be expressed as,

$x^2 + 5x +$ (6)　　　　　$6 = 3 \times 2$

$6 = 6 \times 1$

## Step 2

Now you have to consider the coefficient of the second term. The coefficient should be expressed adding or subtracting the factors that you obtained in the step 1. If the third term has positive sign, you should add the factors. On the other hand, if the third term has negative sign, you have to subtract them.

The expression is,

$$x^2 + 5x + 6$$

Here coefficient of the second term is 5, and the sign of the third term is positive.

In the first step, we expressed 6 as,

$$6 = 3 \times 2$$

$$6 = 6 \times 1$$

Now you should apply the basic mathematics.

You can make 5 adding the first pair of factors of 6.

$$3 + 2 = 5$$

We added the factors in order to make 5 because the third term has positive sign. [So here we should not apply subtraction to make 5 (6 - 1 = 5).]

**Step 3**

Now you are ready to factorize the expression.

$$x^2 + 5x + 6$$

$$= x^2 + (3 + 2)x + 6$$

$$= x^2 + 3x + 2x + 6$$

$$= x(x + 3) + 2(x + 3)$$

$$= (x + 3)(x + 2)$$

Finally we factorized the expression $x^2 + 5x + 6$ as,

$$(x + 3)(x + 2)$$

Let's depict some more examples.

**Example:**

Factorize:

$$x^2 + 5x - 6$$

Solution:

$$x^2 + 5x - 6$$

Here the third term is 6, and it has negative sign.

We may represent 6 as the product of pair of factors as,

$$6 = 3 \times 2$$

$$6 = 6 \times 1$$

The coefficient of x in the second term is 5.

We have to make 5 by subtracting one of the pairs of factors. Here we are subtracting the factors because the sign of the third term is negative.

$$6 - 1 = 5$$

Now we are able to factorize the expression.

$$x^2 + 5x - 6$$
$$= x^2 + (6 - 1)x - 6$$
$$= x^2 + 6x - x - 6$$
$$= x(x + 6) - (x + 6)$$
$$= (x + 6)(x - 1) \quad \text{[Answer]}$$

**Example:**

Factorize:

$$x^2 - 7x - 18$$

Solution:

Here the third term is 18. 18 can be expressed as the product of some pairs of factors.

$$18 = 1×18$$
$$18 = 2×9$$
$$18 = 3×6$$

Now consider the second term, which is 7. And since the sign of 18 is negative, we have to make 7 subtracting one of the pairs of factors.

$$9 - 2 = 7$$

Thus the given expression can be factorized as the following way.

$$x^2 - 7x - 18$$
$$= x^2 - (9 - 2)x - 18$$
$$= x^2 - 9x + 2x - 18$$
$$= x(x - 9) + 2(x - 9)$$
$$= (x - 9)(x + 2) \quad \text{[Answer]}$$

The previous examples were associated with coefficient 1 in their first terms. But how do we factorize the expressions where the first terms have coefficient greater than 1?

The generalized forms of such expressions are,

$$ax^2 + bx + c$$
$$\text{and}$$
$$ax^2 + bxy + cy^2$$

Where a, b, and c are integers, and x and y are unknown terms.

The method of such factorization is quite similar to the previous method. But in this case, you have to consider the product of coefficients of first and third term.

Let's understand the method discussing an example.

Factorize:

$$3x^2 - 5xy - 2y^2$$

**Step 1**
Multiply the coefficients of the first and the third term including their signs.

$$(+3) \times (-2)$$
$$= -6$$

The product is negative which is -6.

**Step 2**
Then express the obtained number (neglecting its negative sign) as the products of pair of factors.
Since the number is 6 (neglecting its negative sign), we may express 6 as,

$$6 = 3 \times 2$$

$$6 = 6 \times 1$$

## Step 3

Now you should make the coefficient of second term by adding or subtracting one of the factor pairs.

When the sign of the product (which you obtained in step 1) is positive, you have to add the factors. And when the sign of product is negative, you have to subtract the factors.

For the following expression,

$$3x^2 - 5xy - 2y^2$$

You multiplied $(+3)$ and $(-2)$, and obtained $-6$ as the product in the step 1.

Since the product is negative, you have to subtract the factors.

The coefficient of second term is 5, so you have to make 5 subtracting one of the factor pairs.

$$6 - 1 = 5$$

## Step 4

Now you are able to factorize the given expression.

$$3x^2 - 5xy - 2y^2$$

$$= 3x^2 - (6 - 1)xy - 2y^2$$

$$= 3x^2 - 6xy + xy - 2y^2$$

$$= 3x(x - 2y) + y(x - 2y)$$

$$= (x - 2y)(3x + y) \quad \text{[Answer]}$$

Now let's discuss some examples:

**Example:**

Factorize:
$$3x^2 + 8xy + 4y^2$$

Solution:
$$3x^2 + 8xy + 4y^2$$

Multiplying coefficients of the first term and the third term, we get,
$$(+3) \times (+4) = 12$$

We may split the coefficient of the middle term as,
$$8 = 6 + 2$$
Because,
$$(+6) \times (+2) = 12$$

Thus we factorize the expression as,
$$3x^2 + 8xy + 4y^2$$
$$= 3x^2 + (6 + 2)xy + 4y^2$$
$$= 3x^2 + 6xy + 2xy + 4y^2$$
$$= 3x(x + 2y) + 2y(x + 2y)$$
$$= (x + 2y)(3x + 2y) \quad \text{[Answer]}$$

**Example:**

Factorize:
$$9a^2 - 3ab - 2b^2$$

Solution:
$$9a^2 - 3ab - 2b^2$$
$$= 9a^2 - (6 - 3)ab - 2b^2$$
$$= 9a^2 - 6ab + 3ab - 2b^2$$
$$= 3a(3a - 2b) + b(3a - 2b)$$
$$= (3a - 2b)(3a + b) \quad \text{[Answer]}$$

Did you notice how we split the middle term in this case?

1) Factorize:

 (a) $ax + a$

 (b) $3a + 12$

 (c) $x^2 + x$

 (d) $5x^3 - 15x^2y$

 (e) $ax + 3x + ay + 3y$

 (f) $ab + a + b + 1$

 (g) $a^2 - ab - ac + bc$

2) Factorize:

 (a) $x^2 - a^2$

 (b) $m^2 - 1$

 (c) $x^4 - 16$

 (d) $81a^4 - 1$

 (e) $9 - b^2y^2$

3) Factorize:

 (a) $a^2 + 2ab + b^2 - c^2$

 (b) $a^2 - x^2 + 2xy - y^2$

 (c) $x^4 + 64$

 (d) $x^4 + 4$

 (e) $x^4 + 3x^2 + 4$

 (f) $4a^4 + 81x^4$

 (g) $9a^2 + 12ab - 5b^2$

4) Factorize:

    (a)   $x^3 + a^3$

    (b)   $8a^3 - 27b^3$

    (c)   $a^6 - b^6$

    (d)   $x^3 + 27$

5) Factorize:

    (a)   $x^2 + 4x + 3$

    (b)   $a^2 + 9a + 14$

    (c)   $x^2 - x - 30$

    (d)   $a^2 + a - 20$

    (e)   $x^2 + 11x + 24$

6) Factorize:

    (a)   $2x^2 + 7x + 3$

    (b)   $6a^2 - 11a + 4$

    (c)   $4a^2 - 12a + 5$

    (d)   $-4x^2 + 8x - 3$

    (e)   $2m^2 + 3mn - 9n^2$

    (f)   $8x^2 + 10x - 3$

    (g)   $2x^2 + x - 6$

    (h)   $3a^2 + 7a - 6$

    (i)   $ax^2 + (a + b)x + b$

# My Notes:

# My Notes: